Towards
ZERO
Waste

How to live a circular life

Féidhlim Harty

Permanent Publications

Published by
Permanent Publications
Hyden House Ltd
The Sustainability Centre
East Meon
Hampshire GU32 1HR
United Kingdom
Tel: 01730 823 311
           International code: +44 (0)
Email: enquiries@permaculture.co.uk
Web: www.permanentpublications.co.uk

Distributed in North America by
Chelsea Green Publishing Company, PO Box 428, White River Junction, VT
05001 www.chelseagreen.com

© Féidhlim Harty, 2019

The right of Féidhlim Harty to be identified as the author of this work as been
asserted by him in accordance with the Copyrights, Designs and Patents Act 1998

Designed by Rozie Apps, Permanent Publications and Permaculture Magazine

Cover design by Gail Harland

Printed in the UK by Bell & Bain, Thornliebank, Glasgow

All paper from FSC certified mixed sources

The Forest Stewardship Council (FSC) is a non-profit
international organisation established to promote the
responsible management of the world's forests

Products carrying the FSC label are independently
certified to assure consumers that they come from
forests that are managed to meet the social, economic
and ecological needs of present and future generations

Towards Zero Waste is a fully revised and updated edition of *Get Rid of Your Bin*,
first published by Mercier Press, Cork © Féidhlim Harty, 2009

ISBN 978 1 85623 328 6

Praise for the book

Towards Zero Waste is my new bible! Whether you've just started your journey to living with less plastic or have been a zero-waste geek for years, this book is a treasure. Packed full of well-researched and well-lived advice it's taught me new things, honed my commitment and – most importantly – inspired me to act!

Natalie Fee, author of *How to Save the World for Free* and founder of City to Sea

Whether you have just begun your endeavor towards a more zero waste life or you have been at it for years and are looking to step up your game, this is the book for you. *Towards Zero Waste* has the ability to help us all live in better alignment with our earth and to live with greater health, happiness, and purpose.

Rob Greenfield, activist and author of *Dude Making a Difference*

This is a great handbook for everyone going into the future. A lot of people don't know where to start with becoming waste neutral and this is a detailed step-by-step guide how to change every area of your life. If everyone read this book and implemented the steps the future would look very bright.

Fergal Smith, heart warrior, surfer and farm manager of Moy Hill CSA farm

Excellent book – informative, encouraging and enabling. 'Despair isn't a good ingredient in the soup that is change,' the author reminds us. That sentence encouraged me, an often struggling environmentalist, to go deeper into this very helpful and easy to read book.

Fran Brady, Eco-Quaker Representative on the Eco-Congregation Ireland Committee

This new book by Féidhlim Harty is all you need to improve your practice and journey in Zero Waste living. Bring it into your home, school, work place or anywhere you interact. Then open it on any page and try to take the thoughts and information into a new step on your journey. There are inspiring examples from the community that buying this book makes you a part of and so Féidhlim encourages you to share your steps with this growing community. It's clear he understands the challenge and yet this book helps avoid shirking our collective responsibility to keep trying, as if our lives depend on it.

Suzie Cahn, co-founder of Carraig Dúlra Permaculture Farm

I think there's a real hunger now to move beyond the basics of recycling and in to a complete overhaul of our lifestyles to stop waste entirely in our food, energy and transport. The idea of being able to live circularly is compelling and imperative. This is an important book by a passionate and informed pioneer of the Zero Waste movement.

Michael Kelly, founder of Grow It Yourself, https://giy.ie

Get ready to plunge yourself far beyond 'waste-deep' into the world of waste-free living. This book is boot camp for waste-warriors wanting to learn the alchemy of incorporating usefulness and fertility back into every discarded morsel in order to live a more ethical, connected and fulfilling life.

Charlie MGee, activist and musician, Formidable Vegetable

This book summarises remarkably the impacts of our wasteful consumption! It is a powerful source of inspiration on your journey towards zero waste, an invitation to create change and to become a mindful consumer. It should be kept to hand to access its goldmine of information at any time, for the love of planet earth and all living beings.

Tristram Stuart, author of *Waste: Uncovering the Global Food Scandal* and founder of ecological charity, Feedback

How refreshing to read an environmental book that is straightforward, practical and not preachy.

Tristan Gooley, author and natural navigator

Towards Zero Waste is an enjoyable book filled with practical ideas for reducing waste in your home and beyond. Féidhlim and his family have worked towards reducing waste for over 20 years and he explains changes we can make in all areas of life in an entertaining, non-judgemental and realistic way.

Stephanie Hafferty, no dig gardener and author of *The Creative Kitchen: Seasonal Plant Based Recipes for Meals, Drinks, Garden and Self Care*

I was inspired by this book; I love the mix of practical everyday ideas and tips alongside bigger picture campaigning ideas. Harty asks us to stop and think about how our consumption creates waste, and does this in a way that motivates us towards change rather than dragging us into despair. Fun, informative and inspirational.

Saffia Farr, editor of *Juno Magazine*

Towards Zero Waste is a comprehensive account of what it means to live a zero waste life and why it's important – with nice details of his own life in rural Ireland mixed with research and statistics of the impact of our current patterns of over-consumption. The author goes beyond talking about a 'circular economy' to a focus on using less and living more mindfully and joyfully with many practical examples and ideas. At a time when many public authorities are using the term 'zero waste' to mean 'zero waste to landfill', and letting in incineration by the back door, this is a really helpful approach. As he reminds us, recycling and 'down-cycling' – where materials are turned into a lower-grade product – no substitute for reducing consumption.

Sophie Unwin, founder of the 'remakery' concept, www.remade.network

The problem of our waste is reaching crisis point: the destruction of the environment through climate breakdown, pollution and loss of biodiversity is an unfolding disaster. Many people are trying to reduce both their consumption and their waste, and this wonderful book gives us both the inspiration to start and practical steps to take. It is a well-researched, easy to follow guide which will be an invaluable companion for everyone who wants move towards zero waste in every part of their life.

Andrew Orr, chair of EcoCongregation Ireland

Much of what Féidhlim has to say is obvious when you think about it, but in our busy lives it can sometimes take a prompt like his to make us think about it. His examples are drawn from real life domestic situations, the familiar problems, and sources of angst, that we all encounter at a practical level. What I like is Féidhlim's honesty; he'll readily admit the things he's had to change and the mistakes he's made along the way. I took from this an encouragement that it's OK if you're muddling through; the important thing is to be conscious of the waste you're generating, even inadvertently, and willing to make changes.

Laura Pardoe, natural skincare advocate and author of *Vital Skincare: Natural Healthy Skin in Just 5 Minutes a Day*

Acknowledgements

As I finish the process of putting together this manuscript I'm struck most of all by gratitude for the many people throughout the world working diligently and passionately in their own areas of expertise and interest. Our society is moving at a tremendous pace towards greater awareness of the reach and extent of our ecological impacts. Hand in hand with this is a steady movement towards genuine solutions at every level of society. So even as the global environmental situation seems to be growing ever more precarious, people in their millions are rising to the challenge.

Every campaigner, every letter to governments or the papers, every business that seeks to offer the very best ecological alternatives to the status quo of ubiquitous plastic, everyone who works in any way to bring about a better world; these are the people to whom I am most grateful his morning.

More specifically, thanks to Mercier Press in Cork for publishing *Get Rid of Your Bin* 10 years ago. In overhauling the book for this edition, I want to express my gratitude to Maddy and Tim Harland for their enthusiasm and their willingness to take me under the warm wings of Permanent Publications and to thank Rozie Apps for her editorial diligence, patience and encouragement.

As part of the research process, I put questions out to our local zero waste group in Ennis and to groups and forums around Ireland. The answers that I received have helped me to get a glimpse into many more bins than just my own. Thanks to you all for flagging problem wastes and the solutions that you have adopted. In a similar vein, I want to thank friends and family around Europe who gave me an insight into bin collection services provided further afield. (Who knew that bin collection is usually free?)

Many many thanks to those who have written testimonials for the book. It is a wonderful feeling to read over your kind words and see it afresh through your eyes. Thank you.

I wish to extend my gratitude to Lynn Finnegan and Róisín Nic Cóil, who each shared just the right information with me at just the right time. Serendipity in action. In our own move towards zero waste, my family and I are assisted in no small measure by Tina, Roy and their family (Imogene and Cheyenne in particular) who run Meanwell Biomarket and Wholefoods in Ennis. I'm grateful for the light you all shine here in Ennis.

My daughters Susie and Kate have helped out in different ways, with both artistic input and assistance with cross-checks. Thanks a million to you both.

Finally I want to thank Elinor, who has carried out much of the practical research for this book and is the careful pair of eyes watching over my shoulder to make sure that the writing is clear, fair and accurate. I'm grateful to have you in my life.

Contents

Dedication

For Elinor,
the practical face of zero waste in our household

About the Author

Féidhlim Harty is a writer and director of FH Wetland Systems ltd., an environmental consultancy company specialising in reed bed system design and other eco-friendly sewage options. His 2009 book, *Get Rid of Your Bin*, was an optimistic attempt to subvert the Irish government's plans to build new incinerators by denying them the waste they need as fuel. This book you hold in your hands is a revised and updated edition, 10 years on from the original publication.

Septic Tank Options and Alternatives followed in 2014 to promote the uptake of effective low-carbon sewage treatment; and *Permaculture Guide to Reed Beds* (2017) offers practical, how-to guidance for DIY reed beds and constructed wetland systems.

Zero waste is part of a general passion for ecological restoration and regeneration which Féidhlim brings to his work, the local community, government policy advocacy and teaching around Ireland.

He lives in Ennis, Co. Clare, Ireland with his family, vegetable garden, polytunnel and a lot of willow trees.

www.wetlandsystems.ie

Chapter 1
Introduction

Humanity has a long history with waste. Archaeologists have delved into the past by examining shell middens, old dumps essentially, where seashells were piled up after the shellfish had been eaten and enjoyed by our distant ancestors. Sites such as these provide a wealth of information about how people lived, what tools we used, what our diets were, and most famously, how we were treated when we died.

I'm not entirely sure that there will be complimentary comments by whoever has the unfortunate job of trawling through our present-day landfill sites for clues about the habits and wastes of our current age.

When my first book, *Get Rid of Your Bin*, was published in 2009, the aim was to encourage household waste minimisation for a variety of reasons. First and foremost in my own mind was to reduce the need for a national municipal waste incineration plan in Ireland. The turn of the century was a time when new landfill sites were being opened and incineration being proposed as a solution. In many ways burning clean paper and timber waste in a municipal incinerator and producing heat and power for the local community makes a lot of sense. However, burning chlorinated plastics, such as PVC, produces very harmful dioxins. Alongside that, we have neither the oil reserves nor the climate resilience to keep making disposable plastics to fuel our incinerators.

My rationale was that by weaning the country off waste, one bin at a time, we could negate the need for new landfill sites and incinerators. Alas, the amount of waste we produce as a country, and as a culture, continues to grow. The heat has been steadily turned up on the incineration process and the first new units are in use here, with more in the pipeline.

Ten years later the speed with which we convert natural habitats into disposable products and packaging is increasing rather than falling. The problems of excess waste are anything but waning. On the contrary there are additional issues that have surfaced in the public consciousness in recent years.

One is the impact of plastics in the oceans; from microbeads to vast quantities of fishing nets lost or abandoned at sea. Another is the growing awareness of our changing climate and the fact that we need to leave 80% of known fossil fuel reserves in the ground to stay anywhere close to meeting our 2°C temperature limits. The 2018 Intergovernmental Panel on Climate Change (IPCC) report[1] says that 1.5°C of global increase will have widespread and serious impacts, so please let's not even consider 2°C if we want a future worth living in (paraphrased). Local communities around the world are struggling to address the challenges of new incinerators, fracking operations, pipelines, gas terminals, and all the other signs of business as usual.

By taking care with how we shop and what we buy, and by buying less overall we can reduce the amount of oil and gas needed for making plastics and carting things around the world. Unfolding around us are oil wars in the Middle East; intimidation and murder of defenders of local lands in Asia, Africa and the Americas; political inaction to address the root causes of climate breakdown; relentless extraction of dirty fuels in a global game of climate roulette; and the current tragedy of the sixth mass extinction on our beautiful blue planet. In a small way *Towards Zero Waste* can help us take positive action and say: "not in our name!"

By being mindful with our purchasing habits and in our lifestyles we can create the world we want to see. We can cut out waste, build rich soil from compost and enjoy meeting local growers and producers. We can create the conditions for abundant, equitable, friendly and beautiful communities. We can hold hands together within and between those communities to build a world that we can all be proud of. This is a vision for the future that we can work towards. Let's step forward and see how.

Start with the Essentials

1. **Practice mindfulness when shopping**. Reduce what you buy. Allow into your home only those things that can leave it again without going via the bin.
2. **Reuse: Return, Repair, Reroute**. Use returnable containers where possible. Repair stuff that's broken. Reroute unwanted but functional things to friends, relations, charity/thrift shops, bookshops, charities or other places that can find a genuine use for them.
3. **Make compost** from all your kitchen and garden food and plant waste. There are some exceptions depending on your circumstances, but basically everything organic in nature can be composted in one way or another.
4. **Recycle as a last resort**. Recycling isn't all it's cracked up to be. We tend to export this dirty industry to countries where protections for people and the natural world are lax. Keep recycling to a minimum, as a solution for what remains after all other measures have been adopted.
5. **Give feedback, push for change**. Changes for the better throughout history, be they great or small, have come about only through people getting out and demanding it. Hand in hand with taking measures in our own homes towards zero waste, society needs help to embrace changes as well.

Some Tools for the Journey

Information on composting, waste minimisation, recycling and simplicity is all readily available on the internet, in the library, from environmental groups, blogs, books and other sources. Use it to complement this book wherever you come up against a stumbling block and have difficulty achieving zero waste. For resources and links visit: www.wetlandsystems. ie/wastetips.html

Reach out to others. Find your local farmers' market and shop there for as much organic produce as you can. See if there is a refill shop in your area or buy in bulk from your local health food shop. Talk with neighbours and see what they're doing to achieve zero waste. Conversations about zero waste are easy now compared with 10 years ago. Go on and find out for yourself.

Grow your own organic vegetables and fruit. Even a 1.2m x 1.2m raised bed can keep you in some vegetables all year round and use all the compost you generate (see the square foot garden plan at the end of Chapter 7). Food from the garden doesn't come in plastic.

Cook from scratch. It will be remarkably difficult to achieve anywhere close to zero waste if you rely on supermarkets and pre-prepared meals. If cooking isn't your strong point, then learn just a few easy recipes. Even something as simple as Farmers' Market Soup is a good introduction to the world of real, packaging-free food and real flavour. Recipe: shop for loose, seasonal veg with your wicker basket or cotton bag; wash and chop; simmer in a good bone broth; blend and eat.

Cancel your bin collection, if you have one. Nothing focuses the mind like burning your bridges. But do so from a place of informed confidence, not blind ignorance. Read on, make your decision and cancel your bin collection as soon as you are ready. You will most likely have some waste at the end of the year, so keep the bin itself and use it as storage for what little you still generate. If you live in a country with bin collection fees, then a trip to the local landfill site will be a lot more cost effective than your annual charges.

Live simply. Re-assess your needs in light of zero waste and environmental impact generally. If you can get another year out of your car, your laptop, your polytunnel – that's another year that the resources stay in the ground rather than being used for producing the next new thing. If you can do without them altogether – so much the better.

Head starts

○ If you have a **garden for composting** kitchen organic waste, and for using the compost you make, you have an immediate head start on the waste reduction stakes.

○ If you have **pets or hens**, to eat meat or kitchen scraps respectively, that is another bonus. Instead of paying for disposal of organic waste that gets smelly in the landfill site and generates methane and landfill leachate, you end up with the satisfied companionship of a pet or with fresh eggs for breakfast.

○ If you have **like-minded neighbours** who would share your bin, you can halve your bin charges overnight and collectively work towards zero waste over time. If you live in an area without charges perhaps you could ask the local council to put the resources they would have invested in bin collection back into a local community garden instead.

○ If you have a shed or shelf space for extra recyclables storage you have the advantage of being able to do occasional recycling drops to the local bring-site instead of paying for collection. Remember that it concentrates the mind when shopping if you have to cart it all away yourself again to the recycling centre.

○ If you live in a town or city with a refill shop, that's a great bonus in your zero waste endeavours. Most good health food shops will sell goods in bulk, which usually reduces both costs and plastic.

These may be a bonus, but all you really need is the desire to get going towards zero waste. Whatever head starts you may or may not have, use good planning and some imagination to make the whole process as simple and straightforward as possible. To ensure success, the new set-up needs to be workable, long term; day in, day out.

How Far to Go

If you look around you, everyone you see has a different reason to aim for zero waste and a different degree to which they are prepared to go. If you choose only what is right for you, the chances are that your efforts towards waste reduction will be more satisfactory, more entertaining (yes, I consider it a fun challenge!) and ultimately more long lasting.

You might want to emulate Bea Johnson, the queen bee of zero waste, and measure your annual waste in a mason jar. Or perhaps splitting the cost of the bin with a few neighbours will be enough, helping to build community, reduce the waste from all the households involved and still keep the process manageable.

How far you go along the path towards zero waste is up to you. Whatever the distance, this book will give you pointers along the way to help reach your destination. Happy travels.

Chapter 2

Why Zero Waste?

Where to start? Anywhere and everywhere really. Plastic is showing up in every corner of the world. It's in our seas in vast quantities, in our freshwater, in the ground, in birds, mammals and even showing up in insects at the very bottom of the food chain. We can be part of the solution rather than part of the problem of plastics everywhere.

Another side to waste, whether plastic or otherwise, is the amount of energy it takes to make it all. Oil and gas are the building blocks not only of the plastics we make, but the fertilisers we use for growing our food and the pesticides we use to kill anything that might try to eat that food before it gets to our plates. Despite the many developments in solar and other renewable energies, oil and gas are the main fuels used for getting stuff (and ourselves) around the world. The more we buy, the more oil and gas we convert into greenhouse gas. For anyone who has been on a media-fast for a whole generation, climate breakdown isn't a good thing.

If you've noticed the weather recently you'll realise that it's already getting unnervingly predictably unpredictable.

For as many generations as we have been around as a distinct species, humans have survived perfectly well without plastics. We've survived without plastic bags, plastic boxes, plastic clothes, plastic shoes, plastic knives and forks, plastic straws, and even, heaven forbid, plastic wrappers around already perfectly well wrapped bananas! Now, because of the sheer volume of stuff we use all the time, for the first time in our collective history we're looking at the stark reality of possibly not surviving for very much longer...

Our culture is built around buying stuff. Keeping the economy going is the dominant story of our time. Any good story has a beginning, a middle and an end. In the story of economic growth it all looked promising in the beginning. Sailing between continents brought untold wealth to the winners, although usually death and disaster to the losers on both sides of the water. Then the idea of trickle-down took hold, the idea that we would all grow rich if the very richest of the rich grew richer. So we worked on that for a couple of generations. But it hasn't worked out quite as planned.

Now it looks very much as if we're in the middle of our story. This is the dark part, where the ending is uncertain and the heroine is flailing about without any hope of reaching her goal and the fate of the whole known world rests on her shoulders. In order to write a happy ending we're going to need to pull together and change the rules of the game. We can't follow the old rules and not suffer the consequences that are all too clearly laid out before us.

If we succeed in gobbling up every habitat, every wild space and every wild thing within it, every drop of fresh water and bit of living healthy soil – what will be left?

Our present course is with disaster. But the beauty of trajectories is that they can change. We can change them. We can make different decisions. We can pursue different priorities. We can step back from the current cultural norms of consumption and economic growth and pursue voluntary simplicity, carefully meeting our needs and enjoying life more along the way.

If you're reading this book you probably have a dozen or more reasons of your own to go for zero waste in your household. I'd much rather focus on the opportunities that are available to help you do that than to get stuck in the nitty gritty of the whys and wherefores, which are pretty depressing. Despair isn't a good ingredient in the soup that is change, but if you want a bit more background then finish out this chapter. Otherwise skip directly to the practical stuff to see how to proceed with making the changes.

The case for zero waste can be loosely categorised under several headings:

- ○ Clean water and oceans
- ○ Protecting other habitats
- ○ Clean air
- ○ Resource conservation
- ○ Saving energy
- ○ Health
- ○ Respect for one another

Although these categories are listed separately, they are all inextricably interlinked. The full implications of our waste load on the world are simply too numerous and intertwined to outline here, but here is a brief overview of some of the challenges.

Clean water and oceans

Plastic waste seems to get everywhere, and cause untold damage as it does so. Ocean gyres pull sea-borne plastics into a large swirl of material that has been dubbed the Great Pacific Garbage Patch. These gyres are testament to the durability of plastics. We're still only learning of the many ways that plastics damage marine life. Ghost fishing gear can continue to catch fish as well as dolphins, turtles and other marine animals for many years after they have been discarded or lost at sea. Plastic bags are eaten by turtles, who mistake them for jellyfish. Rings of any sort, whether bottle seals or six-pack holders, can injure or kill fish and sea birds. Smaller bits of plastic are mistaken for food by all kinds of marine life, edging their way up the food chain as they go. Even whales have been found washed up on beaches, starved, but with stomachs full of plastic.

On the other end of the scale, microplastics are another whole challenge. Larger plastic bits can at least be gathered in great ocean clean-up projects as funding comes on stream. However, these smaller bits, composed of fibres from artificial fibre clothing, tyre bits from road runoff water, microbeads in cleaners and cosmetics, and broken down bits of larger plastic items, are much more difficult to recall. These are so small that they are being found not only in tiny ocean life, but also in drinking water supplies[2] all around the world.

But it's not only plastic that causes trouble in our water. On land, landfill leachate is highly polluting where it escapes into groundwater, rivers or streams, and often contains highly toxic elements. Over the years, solvents, chemicals, batteries and other toxins have made their way into landfill sites and are still contributing to the leachate coming out. Land contaminated with some wastes can be unusable for generations. Although modern landfill sites are much more tightly controlled and regulated than in the past, making stuff and dumping it is simply too wasteful of resources to continue as a way of life.

Poorly treated sewage and industrial effluents make their way to seas on an ongoing basis. This not only impacts on rivers, lakes and seas, but is a monumental waste of fresh water and valuable nutrients that could otherwise be recouped for fertiliser.

Protecting other habitats

Habitat destruction and degradation is another reason to minimise our waste and our consumption. We seem to be in an age of excess, be it excess fishing, logging, mining or whatever. When resources are over-exploited, there are impacts on habitats, flora and fauna.

The world's waste mountain is just the visible tip of resource overuse. Forests are being cut at a staggering rate for grazing land, agriculture and timber. Oil extraction and transportation put coastal habitats at tremendous risk and degrade habitats on land where spillages occur. Fracking has made gas into the newest dirty fuel, with large amounts of fracked gas leaking into the atmosphere during extraction, adding vastly to the generation of greenhouse gases. The fracking process itself is also highly damaging to rivers and lakes and to the groundwater in the areas unfortunate enough to have shale gas in the ground. Oil and gas are the raw materials in plastic, the means of transporting it around the globe and even the fuel that powers the disposal and recycling infrastructure. The more things we buy, the more fuel gets used in the processes involved.

Quite aside from oil, gas and the impacts on climate – toxins are another challenge for habitats and wildlife. Many aspects of manufacture and disposal are hazardous for the natural world. Industrial processes are often highly polluting, or require vast resources to keep up with our diet for ever more of everything.

Physical space is another issue for wildlife. The more land we devote to making and getting rid of things, the less space there is for nature. This is the major problem with biofuels for instance. To create supposedly safer alternatives to fossil fuels we take vast swathes of land out of food production and pristine habitat. But for what? To squander that energy on more plastic gadgets and transporting things and food around the world when we could easily provide for our own needs closer to home.

Clean air

Waste-to-energy incineration promised to solve all of our problems and harmlessly convert a mountain of plastic into clean heat and power. Unfortunately, and rather predictably, it's not that simple. Incineration of wastes produces dioxins and other toxic elements, such as mercury, cadmium and lead, not to mention considerable quantities of contaminated ash. Dioxins are produced during waste incineration wherever chlorine compounds, such as PVC, are included (which is almost always).

Apart from the pollution load from incinerators, it's widely acknowledged now that incineration is simply too wasteful of the resources fed into them each day. Heat and power can be provided in other safer ways, and the materials that we are burning need to be recouped for composting, plastic recycling, metal recovery etc.

Air pollution from the manufacturing of products, processing of raw materials and production of energy leads to acid rain as well as climate breakdown. Transport of raw materials, products and wastes consumes fossil fuels and produces sulphur dioxide (SO_2), carbon dioxide (CO_2) and other pollutants.

Resource conservation

Waste and resource use are inextricably linked. Fossil fuels have a limited availability: new supplies of coal, oil and gas are diminishing at a shocking rate, and still we use them as if there were no tomorrow. Those who study peak oil – the point at which supply is outstripped by demand – suggest that we have reached that point already.[3] There are still reserves of less easily accessible fuels, such as tar sands and shale gas, but these too are finite, and much more costly and ecologically damaging to extract. Surely such a resource waste is not only avoidable, but unjustifiable. Yet we waste it every day in the numerous oil derived products and services we buy and move around the globe.

Raw materials such as stone and metals have less serious implications, although they too are being consumed in vast quantities, using large amounts of energy in their processing. Even recycling glass and metals uses energy that is in excess of sustainable amounts.

Water is an often under-appreciated resource and clean water is much more limited than the rainfall rates here on Europe's western shores would have us believe. Excessive wasting or polluting (with our wastes) of this resource leads to shortages, even in the damp green British Isles.

Climate breakdown is also accentuating the extremes of weather conditions, making dry areas drier. Wasteful lifestyles have led to these changes, but we can still reverse climatic trends with immediate concerted action.

Timber as a resource is replaceable, but not on the scale that we expect it to be available to us for our new oak or mahogany sideboards whenever kitchen fashions change; for toilet paper and tissues; for single sided, single use, chlorine-bleached office paper and so on. I am not suggesting we do without loo paper (although we certainly can), just that we use a somewhat more appropriate source of fibre than trees and reduce other paper use as much as possible. Waste and resource use are two sides of the one coin, and we are overspending on a grand scale.

Instead of simply slowing down how quickly we use up the natural world we urgently need to shift our focus to restoration and regeneration of our waters, woodlands, farmlands and the wild lands of the planet.

Saving energy

Energy consumption and its environmental, social, and political impacts are clearer now than ever before. Oil wars appear to be a current by-product of the oil industry. Whether it is ongoing bombing of the Middle East, or murder, arrests or trouble anywhere with oil reserves, it has long been a cause of conflict. The most effective way to prevent this is to endeavour to eliminate our support for the oil industry. It is almost impossible to avoid oil and gas use at present. This is because it is widely used in conventional agriculture, plastics production, manufacturing, synthetic fibres, ingredients in most cosmetics and cleaning chemicals, transport and travel. However, limiting our energy consumption and our consumption of products generally is essential in this context. With a bit of information and perseverance, huge reductions in oil dependency are possible by changing our purchasing habits and lifestyles.

Apart from the social strife it causes, the dirty fuels that are used to generate our energy are a limited resource. Whether there is war over them or not, peak oil appears to be driving prices up steadily. Even if the peak oil experts are wrong and prices do fall in the future, it seems inevitable that they will climb again. At some point oil will become too scarce to extract economically, certainly in the quantities we are used to.

Global weirding is a term sometimes used to describe the unpredictability of climate breakdown. Recent decades have had some of the hottest summers, mildest winters, longest droughts and most severe floods in recorded climatic history. The exact shape of our future climate is unknown. Whether it's prolonged droughts, rising tides or a shifting Gulf Stream, the changes ahead are unlikely to be gentle ones.

Health

How can our health not be impacted by stuff, and the way we make it, transport it, use it and dispose of it? The impacts are at multiple levels. Incineration and household burning of plastic contaminate the air we breathe. Landfill leachate can pollute the water we drink. Contaminants from both air and water accumulate in the food we eat. Plastic wrap used for food storage is often made of PVC (polyvinyl chloride), which can leach toxic chemicals directly into our food. BPA (bisphenol A) is widely used as a plastic coating in most canned foods, drink cans, many children's toys, plastic food containers and many more plastic items. It has been recently reclassified in the EU as an endocrine disruptor,[4] and is banned by the USFDA for use in babies bottles[5] and labelled a chemical of concern by the USEPA.[6] Even till receipts have a BPA coating on them,[7] which can come off on your hands or easily get into the recycling stream and into recycled paper products as well. With such ubiquitous use of carcinogens in our food, air and water is it any wonder that one in two people are now expected to get cancer in their lifetime?[8]

Happily, waste minimisation and health choices often coincide. If we buy carrots loose from our local farmers' market, eggs from a neighbour who wants the boxes back, and grow our own salad greens, we immediately reduce the amount of plastic entering our kitchens, and improve our diet.

If there was a scale of mental and emotional wellbeing that I could pin on my wall, it would read highest after a visit to the farmers' market and lowest after a trip to the supermarket. I've not researched this, but if I was a gambling type, I'd put money on a direct correlation between emotional health and jumping into a zero waste lifestyle within a supportive community.

Respect for one another

Our local and global social responsibilities are becoming increasingly clear. Whether we affect our neighbours across the fence, or neighbouring countries and continents, we can see that waste disposal has a characteristic disregard for boundaries.

Recycling processes have the potential to be highly toxic. Waste electronic equipment and other 'recyclables' are often exported to less affluent countries with lower wages and less stringent environmental, health and safety legislation. Even recyclables such as paper and plastic have the potential to be damaging to health and local wildlife and habitats depending on the processes involved.

Transboundary air pollution causes acid rain, ozone depletion and global climate breakdown. What we burn and dispose of into the air in one place, whether in our hearths, incinerators or industrial processes, affects not only ourselves and our neighbours, but the whole planet.

Addiction to consumerism is as damaging to us all as addiction to drugs or alcohol. With our addiction to buying things (which we then have to dump, remember) we have bought ourselves into a hole of ecological debt. We seem to be willing to put everything on the line for our habit. Health, wealth, wildlife, and the capacity to even sustain our species on the planet. If that isn't addictive behaviour, what is?

Maintaining a steady supply of oil and gas for our endless plastic products, agricultural chemicals and single-occupancy vehicles seems just a tad unethical. Oil wars were already being fought in the 1990s and show no signs of abating. Maybe if we all buy a lot less stuff and slow down a bit, oil will begin to lose its hold on our society. If we are to meet our Paris Agreement obligations we'll need to leave about 80% of all known fossil fuel reserves in the ground anyway. That's simply not happening yet.

On the subject of wars – global military spending reached $1.8 trillion in 2018.[9] Now that's a waste! Just think of the good work that could be done on ocean clean-ups, clean production, green building, public transport infrastructure, tree planting, regenerative agriculture and rewilding with that sort of budget.

Water wars are another real possibility. Industrial and agricultural processes often consume vast quantities of water; often in areas that can least afford to squander it. When we buy nuts or fruit from water starved areas of the world, we're importing the water footprint of those crops as well. It seems a bit mad, writing here in our moist mild climate, to be eating almonds or avocados that have come from drought ridden California or Chile. Careful purchasing can help to reduce the sheer waste of water taking place where it can be least afforded.

Tackling waste isn't a NIMBY (not in my back yard) attitude to stop littering our own back yard. It's about ensuring that the waste and the results of our wasteful practices don't impact on anybody's back yard, anywhere. Our decisions matter. They impact on somebody somewhere. Out of a spirit of love and respect it's time to turn over a new leaf and see how far this zero waste path can lead.

So, whether you want to reduce landfill, tackle litter on your street, stop fracking in your area or protect the oceans, reducing the waste in your home is an active way to minimise your own contribution to the problems you are trying to solve. By letting businesses and governments know how much we care, and pursuing a zero waste life and a zero waste society we might just get out of the mess we're in. That's a pretty clear motivation for change.

How Much is Enough?

This is a question that has been pondered over millennia, and is now more relevant than ever. 'Enough' is something that we are not particularly good at recognising. The ills associated with overconsumption are clear as day when you open your eyes to them, but somehow 'enough' still seems to elude us. How many people in the world are dying from starvation and how many are dying from the diseases of excess? These are two sides of the same coin. Only by identifying our own 'enough' can we move comfortably away from the media mantra of 'more'.

In our consumer society there is, inherent even in the name, the idea that consumption is our main objective. Always the answer given on TV, radio, magazines and newspapers is "buy more; go on more holidays; gather more experiences; spend more money on this, that, or the other". However, we know that more does not lead to contentment. If 'more' is always the answer, surely somewhere along the way we are asking the wrong questions.

The good news is that we're not alone in our zero waste endeavours. It's now a commonplace conversation when shopping. The rate of change is a joy to watch. Zero waste shops popping up; discussion groups; community workshops. Time is limited for plastic waste. Let's extend that to our overconsumption generally and wake up to the joy of simple pleasures. Our consumer culture will only strangle us if we keep trying to squeeze every penny out of it. Once we loosen our grip on our purse and spend a little more in the farmers' market rather than on prepackaged, imported, chemically-grown produce (cheaper only because of massive tax-breaks and subsidies), then we'll begin to discover the heart in our communities and the glow of positive change in all our interactions.

Enough is enough. The time has come to make the changes we need.

Chapter 3

What Happens to our Waste?

There are endless statistics, tonnages and percentages describing the amount of waste, its sources, relative volumes, and changes with time. However, while this is all very interesting, it can be summarised briefly as follows: **we produce too much waste of all types**. Our recycling rate may be increasing but so too is the volume of waste we produce, and recycling alone, without minimising consumption, is too energy intensive a process to be environmentally sustainable. Not only that, but much of our recyclables end up as litter in far flung countries and make their steady way to the oceans...

Waste Management – Where Does it Go at Present?

The three main disposal options that are commonly used in Europe at present are landfilling, incineration and recovery – split roughly equally although varying from country to country. Granted, recovery includes composting and recycling which are an essential part of a zero waste society, but be wary of statistics – where incineration is used as 'waste-to-energy' it too is included in the recovery figures. Let's look briefly at these three categories before we explore some of the alternatives.

Landfill

Landfill is still a popular method for managing our waste mountain in many countries. Personally I favour it above incineration because it makes our waste mountain visible and spurs us on to deal with the generation of waste in the first place. However the liquid runoff, or leachate, is very polluting and toxic. It contains a diverse range of pollutants generated by any number of wastes in the landfill site. Rotting food provides most of the odour and a lot of the organic pollution load. This rich liquid is accompanied by waste from batteries, solvents, paints, cosmetics, cleaners and so on. Unless landfills are fully lined, leachate can cause serious pollution to local groundwater, streams and rivers. Even when lined, the leachate needs to be removed for treatment, and by virtue of the nature of the contents, full treatment is often impossible since many toxic elements remain in the effluent or sludges.

Landfill gas is another potential hazard. It can migrate through the soil and cause problems ranging from odours to explosions. However, it can also be tapped and burned for fuel, as is done in some cities. This is a very constructive method of dealing with what may otherwise become a potential hazard and a potent greenhouse gas.

While landfills have their drawbacks, they have a role to play as we make the move towards a zero waste society. They can potentially be mined for future resource reclamation and the life-span of our existing landfill sites could be extended dramatically if we were to reduce what we dump.

Incineration

Incineration is favoured by many countries in the EU because it can avoid the political headache of opening new landfill sites and can be used to generate energy if designed as a waste-to-energy facility. It also serves to hide the wastes generated from public view and to boost the 'recovery' statistics for a given country, if used for energy generation.

Whether you call incineration 'thermal treatment' or 'waste-to-energy' the process is essentially the same. Both terms are used to put a positive gloss on the burning of potentially recyclable and/or toxic materials. The result is airborne pollution and significant volumes of contaminated ash that still needs to be landfilled. Contaminants such as dioxin are well documented contributors to cancer and other diseases. This is why burning chlorinated plastics in the home (or in incinerators) can be so dangerous.

Incineration could become a much safer process if all chlorinated plastics and other toxics were excluded from the waste stream. This would allow us to recoup heat energy from wastes without the health hazards for the local community. To do this it would be necessary to eliminate PVC (polyvinyl chloride), other chlorine compounds and other toxics from all shop shelves and industrial processes, which is where the chain of exclusion needs to start. PVC is used for window frames, piping, shower curtains, cling film, waterproof clothing and a host of other applications. An obvious step on a personal level is to avoid PVC ourselves, which is sometimes easier said than done, but possible nonetheless.

Once incinerators are built they need to be fed. The incentive for genuine waste reduction vanishes in a puff of stack emissions. Even where heat from incinerators can be used, it doesn't justify the production of an endless stream of plastic and paper waste just to feed them. If we want to be warm we can put on another layer of clothes, move more and get some health-giving exercise, or at least find environmentally sustainable methods of producing heat. We can build our houses with enough insulation and draught proofing to reach 'passive standard', or using energy efficient district heating using willow biomass crops grown on sewage effluent so we help our rivers thrive into the bargain.

Recovery

Recovery accounts for roughly a third of EU waste collected, and it is by a long shot the most favourable of the disposal options discussed here. However it's worth keeping in mind that this includes waste-to-energy incineration and down-cycling.

Down-cycling is the process of reusing a material for something of lower value, such as turning old buildings (concrete demolition waste) into low-grade crushed aggregate; chipping Christmas trees into bark mulch; or shredding drink bottles into sleeping bag padding. You end up with a lower value material rather than recycling it back into the original product, and as such you keep it out of landfill for ever (as in bark mulch) or for a bit longer (as in sleeping bag padding).

While these processes take a waste product out of landfill and convert it into something potentially useful, they essentially support the ongoing production of ever more things from new raw materials. By contrast, recycling a product such as a glass bottle back to a glass bottle may take a lot of energy, but much less so than starting with the raw materials for glass.[10] It also ensures that we meet our needs for glass bottles without endless mining and encroachment into natural habitats. The same is true for metals, paper and some plastics. Full recycling is better than downcycling, but better still to reduce and reuse[11] instead.

Paradoxically when plastic recycling was first introduced into our area, our own household generation of waste plastic increased. This was because it was suddenly possible to get rid of it without landfilling it. That's why in our house we're working now to reduce not only the waste that is destined for landfill or incineration, but also our recyclable wastes.

Another issue with recycling is the potential toxicity in the process itself. Where materials are exported for recycling, particularly waste electronic equipment and other mixed wastes, the potential for causing harm is great. In Africa and Asia, where many European recyclable materials end up, the toxicity of certain wastes can be very damaging to the people involved in the recycling industry.[12] Not entirely surprisingly, this work is all too often carried out by people with no comprehension of the toxicity inherent in what they are doing and limited opportunity to change their working conditions.

The cleaner the raw materials, the greater the potential for producing high quality end products again and again. Recently China announced that it would stop accepting EU plastic waste for recycling, in part because of the poor source separation and thus the difficulty in recycling it effectively. It had become a waste stream rather than a supply of clean raw materials.

Collection of recyclables has become standard now. Glass, paper, card, metals, timber, compostables and (limited) plastics collection facilities operate in most counties. The many different plastics are sometimes sorted into the different types, but more commonly they are lumped together to varying degrees. It's a tricky category for recycling, what with PET bottles, PVC piping, polypropylene packing and HDPE containers, along with mixed or unspecified plastics. Soft plastics such as polyethylene, polypropylene and polystyrene are often excluded from the recycling collection system because they are simply too awkward to recycle or too bulky to transport around easily.

Hand in hand with collection infrastructure has been the development of separation facilities in landfill sites for other items from the waste stream. Items such as batteries and waste electronic equipment, Christmas trees, bulbs, white goods (fridges, freezers etc.), compostables, timber, paints, solvents, oil, construction materials and demolition wastes are now generally kept out of landfill for diversion elsewhere.

With all recycling, the degree of waste sorting generally depends on the collection company employed to gather and process or export it. Certainly the introduction of recycling facilities has been a welcome alternative to landfilling all our wastes, but we have a long way to go.

Recycling has become the watchword in ecological education programmes; 'my bit for the environment'. Yet it ranks well below reduction and reuse, at the bottom of the three R hierarchy. While recycling has the potential to be part of a move towards waste reduction in our homes and our society, as well as contributing towards resource and energy conservation, this will only work if it is viewed as a last resort for waste already in existence while we move towards a regenerative circular society.

Instead of recycling, if all bottles were returnable, with deposit; if all aluminium cans were just eliminated from the supply chain (as was once the case in Denmark for environmental reasons);[13] if all unnecessary plastic was removed from supermarket shelves, then the need for recycling would be greatly reduced. It is the contents that we want, not the packaging.

The ideal is to reduce our generation of waste, and to appropriately reuse what is left, not dispose of it to far flung lands under the banner of recycling.

Chapter 4

Hidden Waste

There is a whole category of waste that we don't see in our homes or day to day lives. This includes food miles, product miles, embodied energy and embodied waste. I've also included a section here on water footprint and on toxic and hazardous wastes because the impact of these is also hidden from view.

Food Miles

Food miles can be defined as 'The total cumulative distance travelled to get food from the field to the table, including the miles travelled by each ingredient in each product or dish'. So I can choose pumpkin seeds from China that have travelled about 5,800 miles or the pumpkin seeds from Turkey that have travelled only 1,800 miles. Of course, if I can grow my own, that will cut the food miles down to nil for that crop. Calculating for a single-ingredient product is easy. For processed foods with different ingredients it's another whole ballgame.[14] Taking just a pot of locally made organic yoghurt, for example, you need to factor in the local milk products used, imported strawberries, imported sugar, starter cultures, travel to the wholesaler, travel to the retailer and finally delivery to the table.

Clearly from an environmental point of view, the lower the food miles, the lower the impact on the natural world. The less food miles a product travels, the less energy, oil and resources are consumed in getting it from field to fork. Predictably the average farmers' market menu will have a much lower carbon footprint than your average supermarket haul.

Product Miles

When we extend the principle of food miles to 'product miles' we become more aware of every purchase of anything, anywhere, not just for food. The important questions become:

O   Where does a product come from? The closer the better.
O   Where do its raw materials come from? The closer to the site of production the better.
O   How many different raw materials and different sources are used? The fewer the better, usually.

The impacts of the energy consumption of food miles and product miles are vast, and are examined below under the heading of Embodied Energy. The main waste implications of food miles and product miles are in the transportation and the packaging. These are waste sources that do not show up in your weekly bin, but still contribute to our national and international waste mountains.

Different transport options each have their different impacts:

Road: although only 20 times the weight of a car, loaded lorries exert much more wear on roads – in fact for heavier lorries this can be up to 150,000 times more wear and tear than cars.[15] The more road freight that moves around, the more our roads need upkeep and maintenance, the more energy used, and waste generated in the process. The general move towards ever more and ever bigger supermarkets has led to ever greater road haulage, as our supermarket warehouses are now effectively the rolling stock on our roads.

In recent years another threat that has emerged from road transport is that well over half of all microplastics are generated from tyres wearing on the roads; over 500,000T/yr.[16] I have no doubt that biodegradable substitutes will be found, but in the meantime the less we drive and buy products that rely on extensive road travel, the more we can be part of the change for the better.

Air: air travel consumes vast quantities of fuel to get from A to B and as such is very wasteful of energy. Whether it is cut flowers, imported salad leaves or anything else, surely it is not worth risking the health of our planet to fly things in from around the globe that we can grow or make closer to home.

Sea: although water transport is one of the more energy efficient means of transporting goods, there are dangers associated with it. Spilled cargo can be polluting and damaging to wildlife. Even spillages of seemingly innocuous products such as plastic pellets can be fatal to birds, fish and sea mammals when mistaken for food.

Local production: this is an obvious solution, and works admirably for many items. Local food production is already enjoying an upsurge in popularity with farmers' markets and the international Slow Food movement.

In addition to the energy implications of travel, copious amounts of packaging are required to transport items intact. Much of the packaging used for transporting products from A to B would not be necessary if the source and end use were closer together. The volume of plastic and cardboard dumped by retailers is vast, before ever the product is placed in the shopping basket. This would be greatly reduced were it not for the need to ship products large distances from source to use. Packaging as advertising, epitomised by Easter eggs, is another whole arena again.

In short, product miles contribute substantially to both embodied energy and embodied waste. Buy local or grow your own to keep your product miles and food miles to a minimum. Extending this idea further, if we buy fruit and veg in season rather than from cold storage, we get fresher, healthier produce, as well as avoiding the need for extensive refrigeration infrastructure in the supply chain. To put the importance of this in context, coolant gases are listed as the number one source of greenhouse gasses globally,[17] so the less we rely on refrigeration to eat out-of-season food the better.

Embodied Energy

This is the energy that is used in the manufacture of a product, including the production and transportation of raw materials, packaging and finished item itself. The lower the embodied energy, the smaller our carbon footprint. Our carbon footprint is the amount of carbon dioxide we produce through energy consumption by all our travel, purchases, and lifestyle choices. In building our homes for example, products like concrete blocks have a much greater embodied energy than timber, because the manufacture of cement consumes huge amounts of energy.

The waste implications of embodied energy include the wastes generated by the oil industry and other fossil fuel industries; electricity generation and energy transport – via cables, lorries or oil tankers. The higher the embodied energy of a product the greater the waste generated by some or all of these other industries. Like the other hidden wastes, this cost is not borne directly at the weekly bin, but in product costs, taxes and in tertiary costs associated with damage to the natural world.

Climate breakdown in particular is very clearly a new normal, with regular reports of forest fires, flooding, drought and extremes of temperature throughout the world. At the time of writing, although the current US president denies that climate breakdown exists, the government there has seen an unprecedented rise in storm damage costs from an average of c.$20bn/yr in the 1980s to a spike of over $300bn in 2017.[18] Regardless of the cost implications, the impact on people and wildlife the world over is heartbreaking. IPCC[1] projections point towards steady deterioration if we continue with business as usual.

Embodied Waste

Embodied waste is the waste generated during the manufacture of a product and its constituent raw materials. As an example, take a 'cardboard' milk carton, before even including the milk: paper, plastic and aluminium are all needed. Effluent from paper bleaching and dying processes can be very polluting. The plastic used in the carton has another set of wastes associated with its manufacture, as does the aluminium foil sheet used to keep out sunlight. Then the dairy industry needs to be considered. Finally there is the shrink wrap plastic used to hold the cartons together for delivery to the retailers. Every product we buy has a similar trail. To avoid making our purchases consciously, would be to avoid taking responsibility for the impacts on our living planet.

Fortunately a very simple solution exists: the most straightforward way to avoid hidden waste is to buy local and in season, grow you own and/ or to minimise what you buy.

Keep in mind that beyond a surprisingly low level of material wealth, more stuff does not buy us more happiness. The basic requirements for living are essential to provide some degree of physical comfort, but beyond that, happiness cannot necessarily be increased by boosting your income or your spending: i.e. by having more stuff.

To listen to the stock market reports on the news you'd think that economic growth is as essential as the air we breathe. For the economic model we've bought into, it probably is, but for every other thing that we hold dear; our planet, our capacity to grow food for ourselves, our families and friends, all forms of wildlife, it's the opposite – that growth that we so diligently pursue for our economies means that we place ever increasing pressure on the world and every living thing within it. The time has well passed to reassess our priorities and joyfully embrace a conscious voluntary simplicity in our lifestyles.

Bear in mind that there are two strands to zero waste. The first is that initial step of reducing the waste that enters your own home. Then there is remaining mindful of the waste incurred in the supply chain as well. We can do a bit by handing back packaging to retailers – and it shows up as less waste for us to deal with at the end of the year – but we do a lot more by selecting products and suppliers that simply don't generate waste in the first place. This generally has spin-off benefits for every other aspect of the natural world and society as well.

Water Footprint

Just like carbon footprint is the measure of the energy used to make a product or process, our water footprint is a measure of the amount of water used to produce a given food or product. In the temperate climate off the west coast of Europe, you'd think that water footprint isn't all that important a consideration when we shop. However, crops such as cotton, avocados, almonds have very high water requirements, and they demand water at the point of production – areas that are already water stressed as it is.

Thus as we tackle this ecological and social justice issue, it may not be shorter showers that are the most effective solution to saving water, but care with what we put in our shopping basket. To follow this example, a 10 minute shower uses about 50-100 litres of water, which is available (here in Co. Clare where I am writing this) in abundance, whereas a kilo of avocados in my shopping basket can require upwards of 2,000 litres of water[19] taken from landscapes that can ill afford to lose it to evapotranspiration.

Toxic and Hazardous Wastes

When it comes to waste minimisation, it makes a lot of sense to pay particular attention to items such as household hazardous waste. The lower the toxic element in our waste the better. This is particularly true from an environmental perspective. Not only is the disposal of these items potentially hazardous, the manufacturing process is also more likely to be toxic to people and wildlife at the place of production.

When we consider the term 'toxic waste', the images are usually of large polluting industries rather than the contents of the shelf under the kitchen sink or in the garden shed. Yet although toxic and hazardous wastes are generated in industry, they end up in our homes and elsewhere whenever we buy them. The regulations for industry are becoming steadily more stringent, but some rules cover home use also. For example, there are many items that are prohibited from disposal via the rubbish bin. These include what is termed household hazardous waste.

Household hazardous waste contains such items as paints, solvents, cleaning fluids, batteries, fluorescent tubes, CFLs and LEDs, biocide containers, medicines etc. They are many and varied, and have the potential for great environmental damage if disposed of inappropriately. Even when used and disposed of as intended, toxic and hazardous materials have a tendency to escape into the wider environment during manufacture, use, recycling and/or disposal.

Traditionally such wastes would have been disposed of in landfill, despite the inadequacies of early landfill lining, where they can escape into the groundwater. Another route of easy entry into groundwater is via the toilet or outdoor gully pipes. This is obviously an inappropriate way to get rid of paints, solvents, leftover biocides, leftover medicines and other hazardous materials. Your friendly septic tank bacteria won't stand a chance against such an onslaught, leading to poor treatment, odours and local groundwater pollution. More commonly antibiotics and other pharmaceuticals end up in the loo (and then our drinking water) via urine. The most effective way to avoid this is find safer ways to maintain your health than using prescription drugs.

By now, most waste collection sites and bring centres have designated household hazardous waste storage facilities for some or all of the following wastes:

○ paints, varnishes, strippers and thinners
○ batteries
○ fluorescent tubes, energy saving light bulbs (CFLs), thermometers (all of which can contain mercury)
○ waste electronics, including LEDs
○ weed killers, insecticides, antifreeze, poisons, fungicides
○ aerosols
○ fertilisers
○ polishes, adhesives, glues, inks, sealants
○ cleaning agents, detergents, bleaches, disinfectants, caustic soda
○ waste oils
○ photographic waste (darkroom chemicals)
○ old medicines, waste cosmetics

While appropriate collection is better than ad hoc disposal, the best option is elimination or minimisation at source. If it is possible to reduce ordinary household waste to a minimum, surely it is possible to eliminate (or greatly minimise) the toxic element within it. For example, if you choose to have battery operated gadgets in the house at all, then make sure that you try to use only rechargeable batteries. In this way the volume for disposal can be greatly reduced. When buying electronic equipment, ask about lead, mercury and other heavy metal content and about the potential for responsible recycling post-use. Better still, reduce new purchases of electronics and other items with hazardous components or ingredients to a minimum. Second hand electronics are basically resource-neutral, so try to both source and sell the equipment you need or have.

Instead of bleaches, abrasives, scouring agents and solvents (a.k.a. household cleaning chemicals) it is possible to get very good results with bread soda (bicarbonate of soda) and vinegar. It's healthier for us and for the natural world around us to use natural cleaners on all kitchen surfaces than to use chemicals that shouldn't even be landfilled or incinerated.

For antibacterial agents a simple mix of tea tree oil and water can be used for most applications. Lavender oil in the mix sweetens the smell and is also antiseptic. I find that the trickiest thing to clean is the glue from the more persistent jars labels. A quarter teaspoon of bread soda (bicarbonate of soda) on a cloth and a little elbow grease works wonders and the jar is ready to use again.

Paints, glues and cosmetics are getting less toxic by the decade. Yet it is by consumers demanding a change to safer products that manufacturers are compelled to find new ways to make the products we want. To hasten this change, carefully source the natural alternatives in your local health food shop and ask your current retailer for natural alternatives to the conventional products. This sends a clear message back to the manufacturers and they will in turn change their methods to keep their customers.

Chapter 5
Zero Waste Solutions

While Reduce, Reuse, Recycle is all well and good as an aspiration; if it were practised and applied to all levels of society this book would not be necessary. Clearly we still have room for improvement. Below is an amended version of the 3 Rs, in order of preference.

1. **Reduce** the intake of items that don't make the zero waste grade and items that have high embodied energy or resources.
2. **Research** the lowest-impact alternatives for the items that you need, but don't meet your zero waste criteria.
3. **Refuse** to take into your home what you cannot easily reuse, repair or reroute again.
4. **Reflect** on where true contentment lies and practice voluntary simplicity.
5. **Reuse** things that can be reused; either yourself or by others.
6. **Return** packaging items for reuse where possible.

7. **Repair** things that are broken but essentially fixable.
8. **Reroute** unwanted things to family, friends or your local community.
9. **Compost** all kitchen and garden waste that is organic in nature.
10. **Recoup** energy from clean paper and clean timber. (Do not burn wood composites, plastics, laminated, coloured or glossy paper or treated wood.)
11. **Recycle** (via the local council collection centre, for making the same thing again, such as glass, metal, card).
12. **Down-cycle** (via the local collection centre, for making into something else such as timber into fibreboard, most plastics into lower grade products etc.).
13. **Request change and offer constructive feedback** to shops and manufacturers that you buy from and to governments whose policies impact on your local area and on the wider world.
14. **Remember** that life isn't perfect. There will inevitably be some waste left over. This can go to landfill and you don't need to feel guilty about it.

These can be summarised as five distinct steps:
- ○ Step 1 – Practice mindfulness when shopping
- ○ Step 2 – Reuse: return, repair, reroute
- ○ Step 3 – Make compost
- ○ Step 4 – Recycle as a last resort
- ○ Step 5 – Give feedback, push for change

If you can satisfy the first three steps, then food miles and product miles notwithstanding, your ecological impact will be minimal. However living in our culture of stuff (it's not called a consumer culture for nothing) there will inevitably be products and packaging entering the home that will not fit into these categories, hence the need for options 4 and 5.

The next section looks at these steps in more detail.

Step 1 – Practice Mindfulness When Shopping

This step is to explore how and where you shop. Most of us shop in the nearest supermarket. They're usually convenient, cheap and consistent. But they've got their drawbacks. A lot of the products come wrapped in plastic for a start. The long distances that supermarket products travel mean that they need robust packaging to weather the journey.

The long food miles have other issues as well. Imported fruit is often harvested under-ripe to arrive looking presentable. It will get to you in one piece, but the quality won't be anywhere close to the ripe, local equivalent. The long haul that food makes to get to the supermarkets also takes a heavy toll on the natural world. The carbon footprint of food miles means that every time we feast on imported fruit or veg, the climate takes a hit.

Every time we spend money, we vote for how we want to world to be. It's that simple actually. It's clear that the way we have been doing things in the world isn't working for our society, climate, health or our shared home. Instead of doing things the usual way, we need to practice mindfulness with our spending habits. Each time you reach for your wallet or purse, ask yourself, what world am I voting for here?

It should be a simple matter to read from a label what sprays have been added to your food and whether the contents or packaging are recycled or recyclable. Alas, there is no legal requirement to state what chemicals are added to your crops or livestock until it reaches the factory as ingredients for processing. Neither is product recycling labelling clear or even helpful. That twirly triangular or circular sign with the three arrows can mean 'recycled', 'partially recycled materials', or 'recyclable' (which in itself can mean virgin-timber-derived, chlorine-bleached paper or card, but which can technically be recycled). The symbol and its many variants are also in common use on materials that are patently not recyclable in practical terms, such as 'cardboard' cartons. As such it's pretty useless at best, and greenwashing at worst.

It can take a lot of research to find out what the impacts of our purchases are, but an easy way to approach it is to find places to shop that simply don't need packaging and food that is chemical-free, so you avoid the embodied waste in the growing practices.

Let's start with a look at the weekly grocery shop. Notice how many things are in there that you do not need or even want. There may be nets for the oranges; plastic around the bread; multiple layers that appear to be necessary to contain pizzas, chocolates, sweets, toiletries and a host of other 'essential items'. All these wrappings are surplus to your requirements, why pay for them twice? Once to buy, once to throw away. Thrice even, if you consider the costs to the natural world. Even items that benefit from containers; such as milk, eggs and flour, could easily be bought in returnable containers if the infrastructure were in place.

Avoiding the supermarket route is the easiest way to sidestep the packaging mountain. Small local suppliers often sell fruit and vegetables loose rather than packaged. Local farmers' markets are becoming ever more common. Suppliers of local produce cannot afford, or do not waste their time and resources, on expensive packaging. Return the egg boxes to your local supplier; buy your milk from a farming neighbour in a reusable container. At least flour comes in a plain paper bag that can be used as a firelighter when you are finished with it, or for lining the compost bowl in the kitchen, so it comes out clean when tipped into your outside compost bin.

Impulse buys are the warning cry of books on saving money. They are often unnecessary and not always even wanted. Just ask the quick question with each purchase: OK, how exactly am I going to dispose of this product/container when I am finished with it? For example, a big plastic picnic bench may have a longer life expectancy than a chocolate wrapper, but it takes up more space in a landfill site when its useful lifespan ends. One of the recommended ways to overcome the draw of advertising and impulse buys is to make a shopping list beforehand and to purchase only what is on that list.

The recommendation here is not penury, it is making conscious choices. Cutting out the extraneous, the unnecessary and the unwanted shouldn't be a difficult thing to espouse. Yet the weight of advertising is geared towards selling people exactly the opposite. Much of what we do in the world is unconscious. We don't mean to alter global weather patterns when we buy our new cars and fridges and a myriad of other consumer items, but unless we reverse current trends in our resource use, that will be the result. By making conscious, informed choices in our lifestyles we can overcome many of the ecological and social challenges that currently face us.

Seeing the connection between what we purchase and what we pay in terms of waste disposal and the wider impact on the world is a matter of re-educating ourselves. Then it becomes easier to change our habits to reflect what we want. If you are health conscious you will know the implications of including sugar and processed foods in your diet. We need to practice health conscious shopping for the whole planet.

At face value it may be difficult to see the difference between Fairtrade goods and ordinary products, yet the contrast is stark for the farmers who grow the cocoa beans, sugar cane and other crops. Our shopping habits echo around the world, either helping to create the world we want, or going against the values we espouse, depending on what we choose to purchase.

Shopping mindfully won't always lead us to the cheapest purchase. In fact the contrary is often the case. Some things will be more expensive, but you will generally save money on your overall shopping basket by buying only what is needed. Shopping with this conscious awareness means that you look for products that you genuinely want to have in your life and products, like Fairtrade, that do more good in the world than harm. Watching the volume of waste is only a springboard to mindfulness about how you spend your money. If money talks, then how you spend yours is a direct way to speak up for a better world.

When you are shopping, remember that the items that go into your basket all need to leave your house again in one way or another. If you fill your basket with things that can be composted, reused, easily maintained and repaired, then you can be sure that you won't be paying for their disposal as well as for their purchase. Try shopping really mindfully for a month and see how it reduces your waste.

Refuse

Stopping our waste intake for each thing in our bin is the aim. Elimination of all waste sources is not really possible, but it is surprising how quickly you discover that such-and-such an essential household item is actually quite dispensable. Once I bought toothpaste that advertised 'No sodium bicarbonate, no surfactants, no chemicals'. I bought it and used half of it. Then I discovered that water, which also has no sodium bicarbonate, no surfactants and no chemicals (fluoridation and chlorination excepted), tasted nicer and did just as good a job.

In these small ways can we gradually refuse ever more waste and move towards the change that is needed both in our households and in the world. The long and short of it is that as a society we produce far too much waste. Some projections put waste generation at an additional 70% globally over the coming three decades.[20] If we want to embrace a regenerative future we need to minimise our waste generation beyond all government targets and projections. Given how easy it is to simply buy stuff and dump it, shrinking this waste stream will take a concerted effort. Fortunately we can make a good start with our own shopping baskets, our own bin and our own household decisions. Millions of others are doing likewise, and together we're already seeing changes take place, which all buck the business-as-usual trends.

Research

OK, so you want shampoo, washing-up liquid and stain remover... your local supermarket gives a dazzling array of plastic containers to choose from. With some research you will be able to find alternatives. Ask friends, check your local health food shops, do some internet research.[21] We can look for ways to do without a particular item, or to source it without plastic. There is a balance though – not much point ordering some indispensable product in a glass jar if it has to be bought online and posted from the far side of the world just to avoid the plastic bag it comes in locally. At that point perhaps the saner choice is to research how to make it or something like it yourself. If you want to be minimalist; do you need it at all? Life seems to become a little simpler when you look it at in these terms.

The world is ready for the shift away from plastic. More and more companies are responding positively. I'm not sure that ten years ago we could buy shampoo bars in our town here in the West of Ireland; Nor get refills of washing-up liquid or laundry detergent. I'm still not sure if we can get stain remover, but I have seen it for sale online as a bar rather than a plastic bottle, which is a good development.

Research also applies to the wider ecological and social impacts of our purchases. Zero waste is one strand of this, but not the only one. Different people will have different priorities here. Many people buy local to minimise food miles, support local jobs, or avoid buying products containing palm oil (due the destruction of the rainforests in Indonesia and Malaysia, the source of over 80% of the global supply).[22] Others avidly avoid Nestlé due to their long history of dubious marketing practices.[23] The BDS movement[24] is urging boycotting of companies, divestment from banks and pension funds etc., and government sanctions to help bring an end to the ongoing violence and oppression of Palestinians. This is essentially the same process that worked successfully against apartheid in South Africa.

Whatever your motivation, it is useful to be well versed in the impact of your purchases. The world is such an interconnected place that when we spend money in one area we are inevitably going to impact on somewhere else. It's important that our impact is as positive as it can be, in order to bring about the world we want to hand on to our children.

Reduce

While researching this book, I asked people who I knew to be dedicated to zero waste what their biggest item of rubbish was. Almost universally the response was soft plastic; the wrapping around bread, cheese, fruit, veg, meat, fish etc. It's not currently recyclable here and is quite difficult to avoid. However, if we bring our zero waste ideals into our shopping basket we can endeavour to refuse the bread with a plastic wrapper, and opt for the one from the market instead. We can bring a plastic box or lidded glass bowl or jar to the fishmonger or butcher and refuse the plastic bag. We can refuse the cheeses that come in plastic and opt for unwrapped ones. There are many ways in which we can exercise our zero waste muscle and reduce our intake of plastic at the point of purchase.

Voluntary simplicity

There is another side to mindful shopping. It's not just about choosing this or that. Sometimes the decision is to simply avoid spending on a particular item in the first place. If we are shopping mindfully, it may be well to ask: is this purchase really necessary? To take an example, the clothing industry makes enough products for each person on the planet to have nearly 14 new items every year.[25] Something is clearly amiss with that. By practising voluntary simplicity we are choosing a life with lower impact and removing our cash from supporting the madness.

Simplicity is much bandied about as a necessity for stress-free living, for a feng shui house, or the ultimate in interior design. However the essence of simplicity is the cultivation of a feeling of enough and a feeling of contentment with life and what it offers. This approach doesn't seem to get much media coverage because it doesn't sell advertising space. Not buying stuff is the ultimate waste minimisation tactic. It also reduces the time you need to spend earning the money to buy things in the first place.

Simplicity hasn't just jumped off the pages of the household and fashion magazines within the last decade. Simplicity in one form or another has been actively pursued by those who value it, since at least the time of the Ancient Greeks. Since then it has appeared in writings or practices from many cultures and traditions throughout the world.

There are many different factors that may prompt someone to pursue simplicity. One person may want financial freedom from a mortgage and may downsize in order to achieve this. Another may want to have space and solitude for writing or painting. The spiritual and contemplative traditions have always had numerous examples of simplicity. For some, the reason may be global consciousness, social justice, care for our common home or a wish to avoid using more than our due. Or it may be to avoid negative impacts on those who make our cheap clothing and electronic equipment and grow our imported food. For others the motivation may be personal health or that of their family.

"To live more simply means to live more purposefully and with a minimum of needless distraction," writes Duane Elgin, author of Voluntary Simplicity.[26] Typically a greater enjoyment from life is derived from such conscious living. More satisfaction can come from what is already available than the constant expectation of possessing the latest advertised thing. Simplicity typically fosters greater self-reliance. There is also a greater sense of being able to choose your own priorities. If you do not take the official party line on how to live a life, you can begin asking yourself who you really are and what you really want to do. Advertising doesn't foster such independent thought.

Consumption patterns vary hugely between different cultures in today's world. Our aggressively consumption-oriented culture has led to vast disparities between rich and poor nations, and the rich and poor in any one country. You are what you buy it seems – but only if you play the game. As soon as you step outside the box, the magical lure of the advertised somehow begins to loosen its grip. Life can once again begin to replace the consumption treadmill.

Don't underestimate the buzz that can be got from buying stuff though. If we're going to wean ourselves off anything addictive we risk an emotional crash afterwards. But the rewards are plentiful. If shopping is or was your thing, then look around to see what other gifts life can offer in order to get a feeling of fulfilment, meaning and contentment in each day.

From a waste minimisation perspective, simplicity of lifestyle has a lot on its side. The less we buy the less we throw away. The smaller the house we live in, the less stuff we need to fill it up. A positive feedback loop is that health, environment and finances tend to go together. Quite the contrary of what may be expected, the cheaper option and environmental option often go hand in hand, and are often healthier too. For example, if instead of driving to the gym, you start cycling to work, you then burn less fossil fuel, spend less on both the gym and petrol and potentially get more frequent and regular exercise too.

While I want to point out the benefits of voluntary simplicity, note that this isn't a guilt trip. The whole point of simplicity is conscious living. It is about mindfully directing your life for yourself, rather than letting television, politicians (or zero waste books) tell you how to live.

Take it at whatever level you wish. If you picked up this book, then moving towards zero waste is important for you. Shop mindfully as a way to achieve that. If you want to take it to another level and slow down to the joy that life offers, then voluntary simplicity is one way to help you get there.

Step 2 – Reuse: Return, Repair, Reroute

When it comes to reducing the amount of waste that ends up in your bin, it makes sense to reuse everything that you possibly can. In this context I am taking a fairly broad definition of reuse, including fixing broken things, using second hand shops, charity/thrift shops, eBay, Buy and Freecycle, LETS (Local Exchange Trading Systems), 'free to a good home' forums or giving away to friends and family.

Returnables

The last decade has seen an increase in people's desire for returnable containers such as milk bottles and glass yoghurt pots. This will become much more common as our society chooses to adopt more regenerative practices and as the legal loop closes around the wasteful way in which we use plastic for single use items.

Remember when every pub had stacks of empty glass bottles out the back? They were all on deposit from the supplier; all scheduled for a good wash and a refill. Crushing these and melting them down for new bottles is much more wasteful of energy.

Ideally we would do well to reintroduce the infrastructure for centralised collection, washing and reuse of bottles and jars. This is the step that our governments and society needs to make on the path to zero waste. It is far better than recycling because it only requires collection and washing to make the product fit for use again, instead of expending vast energy inputs for crushing, melting and remanufacturing the glass.

I imagine and hope that it will be a short time before reuse of glass containers is widespread again. However, in the meantime watch for opportunities to reuse, repair and reroute instead, all the while advocating for changes in infrastructure.

Whether you want milk, shampoo or wash-up liquid, you want the product, not the container. Ask your usual supplier to stock returnables, and to take them back again. Health food shops, local markets or CSA (Community Supported Agriculture) schemes are often great about facilitating you where they can.

Libraries are an excellent example of returnables in action. Some schools offer school book rental schemes. These are just some examples, but there are plenty more. Use charity/thrift shops as a library for clothes, buying what you need there instead of in the high street shops.

We need better infrastructure for returnables on all levels of society, so get good at asking, either in person or by writing a letter or email. The more customers request returnable containers the more they will appear on the shelves. Push for change at government level too. Often the much-needed legislation is already drafted and sitting in an office waiting for enough public pressure to overcome the resistance posed by industry lobbyists.

Repair

If it's broken, fix it. Whether it is car, a laptop, the garden shed or the kitchen shelves, if you fix it up and get it serviceable again, then it won't make its way into the bin. If you consider that up to half of the carbon footprint of your car can be in the initial manufacturing process,[27] then it makes a lot of sense ecologically to keep old ones on the road (and use them less) rather than scrapping them prematurely, even if that's an upgrade to an electric model.

Not everybody has the skills for fixing things, but keep an eye out for repair cafés, mens sheds, knitting and stitching groups or knowledgeable neighbours who can offer assistance or advice.

Many computers, power tools, mobile phones and the like are manufactured with redundancy in mind. Repair simply isn't factored into the design. If I was of a suspicious nature I may even suggest that this was intentionally calculated to sell more products, but I won't. The internet has a host of hacks to get your electronics working again. Some will work, some won't – but if the alternative is dumping the item in question then it's worth a try. There is a lovely feeling of satisfaction from getting gadgets working again yourself.

Rerouting elsewhere

Rerouting applies to anything that you can sell or give away for appropriate reuse elsewhere. Many household 'waste' items are merely objects that have outlived their wantedness in the house and are taking up valuable storage space. Rather than dump your cider-making kit, your old Lego or aunt Fanny's tea set, find a suitable home for it. Try selling it or letting it go to someone who can use it. Check your friends and family too. You never know, maybe your cousins are in need of a sofa or a baby's cot. Baby's clothes are particularly good at doing the rounds.

Cars, bicycles and furniture have long been bought and sold, long after their first owners have forgotten about them. Second hand shops, local ad newspapers and bulletin boards in local shops all help in circulating useful but unwanted items.

The internet has introduced a whole array of opportunities to pass underutilised items on to new homes, either for free or at a profit. Freecycle, local 'free-to-a-good-home' Facebook pages and eBay can all help us find a good home for things we no longer want. This way you can keep things moving around for free and avoid the need for people to buy new stuff. It helps to cut down on clutter, reduce new purchases and associated resource consumption, and minimises the amount of dumping that goes on in your household and wider community. (Amazon keeps second hand stuff on the move too, but I hesitate to even list them here, since they score close to the bottom of the league on climate impact and sustainability.)[28]

Don't dump junk onto people though, or it will go straight into their bin, which isn't exactly the idea. This is not a dumping frenzy on the latest family member with a new house, nor donations of moth eaten blankets to an aid appeal. Nor is it an injunction to dump last year's colours and do a new haul of the haberdasheries. It's a conscious exploration of the contents of your home to ask: "What do I choose to keep in my life and what items can I gracefully release into the world again?"

Reusing things yourself

Having returned packaging to the source for reuse where possible; repaired anything that's broken and rerouted good but unwanted stuff elsewhere, you'll still have things left over. Even if something is beyond reusing as the manufacturer intended, why dump it just because it has outlived its original purpose? We can repurpose yoghurt cartons for freezer containers; old natural material carpet off-cuts for compost heap covers; or timber off-cuts as kindling for starting the fire... and so on ad infinitum.

Autumn is a time in our house when the hoarded bottles and jars either get reused for our own preserves or scooped up from their overflowing summer storage box for use by friends. Using waste containers around the home instead of buying freezer bags and plastic cartons helps us to avoid buying new empty plastic containers and then dumping perfectly serviceable ones that have come up with food in them. However, this sort of reuse is not a cure-all. We will never stem the constant tide of waste unless shift our shopping and retail infrastructure towards items that come in returnable containers.

Waste timber and paper can find a reuse of sorts in your own home by using it as fuel. Reusing light timber for kindling and making firelighters from old newspapers and grill pan fat or candle butts is an excellent way to be creative with what might otherwise be a difficult waste to dispose of.

Warning: be sure to avoid burning wood composites due to the glues present, or wood that has been treated with preservatives, painted or varnished. Also avoid burning glossy or highly coloured paper. Remember that burning plastic waste will introduce an array of toxins into your neighbourhood too.

Making space at home for reuse, repair and rerouting

To make reuse and rerouting easy, it is important that good structures are in place. Every house and every family will have different circumstances in terms of the space that can be dedicated to moving towards a zero waste household. Each household will also have different materials for reuse and rerouting, and hence different destinations. It is important that the system you adopt works for you and your family.

Before you start your waste minimisation drive, the following may be useful:

○   a designated place for items in need of repair
○   a shelf or box for reusable and reroutable items en route to friends, family or second-hand shops etc.
○   a suitable container for kitchen compostables
○   a box for materials en route to recycling (see the recycling section for more details on this)

Good storage is essential for items no longer needed, but too good for recycling. This can hold items en route to friends, family, charity/thrift shops or repair shops. Otherwise these will continually trip you up at the front door until in a fit of exasperation you just fling them into the bin. (That of course, would be contrary to the spirit of the endeavour.)

Sourcing returnables and recycled products

Returnables will only appear on the shop shelves if we ask for them. Try to find returnable, refillable bottles and jars for the items that you need on your weekly shop. Sometimes the easiest way to do this is to start shopping in the country market where every jam pot has obviously been given a new lease of life by the local preserves expert.

The availability of recycling facilities is directly dependant upon the market for the materials. If we buy more recycled paper products and stationery then the demand for waste paper will increase. If we request reusable bottles in the shops then a supplier will start supplying them because a market will be created for them.

If you can't find what you need locally, look for the best supplier within your county or country, or set up a local co-op to get deliveries in bulk for several households.

Step 3 – Make Compost

Let's not beat about the bush here, composting is magic, pure and simple. You take a load of useless food trimmings, gone off bits, garden clippings and mowings that have the potential to rot into a smelly, methane generating mess and instead convert them into dark rich humus, the key ingredient in healthy soil, sequestering carbon from the atmosphere as you do so. It's possibly the most effective way for us to reverse climate breakdown in our own gardens. Of course it will take a lot of us, but that's the thing, there are lots of us. So don't shy away from composting. It's alchemy in action.

In most places it's not permitted to put food waste into either the main bin or the recycling bin, so I'm assuming if you're reading this that your food waste is either being collected in a small brown bin for municipal composting or else you are already composting it yourself to build up the soil in your own garden.

While the growing availability of municipal composting is an excellent development, it does mean that those who use it are simply giving away valuable nutrients and biomass that could otherwise be growing vegetables in their own gardens. Also, if you live in a country with waste collection charges and you decide to cancel your bin service, you'll need to deal with organic wastes somehow. This is best done in your own garden or in a community composting project if you live in an apartment.

If it will rot, it will compost. Apple cores, leftover gravy, old leather boots, cotton duster rags (that were once your favourite shirt), cat litter, newspapers, fish bones, chicken stock pot bits, garden clippings, lawn mowings, sink filter bits, toe nails... The list is a long one. Different items need to be dealt with in different ways to ensure that the process is safe and hygienic, but the long and short of it is that lots more will compost than many people realise.

There is a detailed look at composting in Chapter 7, along with beginner guidelines on growing your own food (the obvious way to make good use of all that homemade compost).

Step 4 – Recycling as a Last Resort

For the stuff that is beyond use, recycling is the next step. Recycling is not the answer to our problems though. For years we have been exporting much of our plastic waste to Asia for recycling there, where laws around employment conditions, child labour and environmental protection are all far more lax than at home. With China's recent decision to halt imports of European recyclables,[29] the contents of those recycling bins begin to look more like what they really are, rubbish.

In the EU we have lumped recycling and recovery together for statistics purposes. This means that plastic and other waste that gets incinerated can be included in the overall recycling/recovery figures. This may be daft enough if the materials were timber, paper and clean card, but it's lunacy when it comes to plastic. Plastics are increasingly being made from fracked gas.[30] The extraction process in fracking releases a shockingly high percentage of gas into the atmosphere, where it has a greenhouse gas effect 34-86 times[31] more potent than CO_2.

At first glance over lifecycle analysis figures, plastic may have a lower embodied energy than glass bottles or even paper bags. However, this does not necessarily take into account the fact that plastic manufacturing has just got a lot dirtier with fracking. Also, we're pretty far down the list of zero waste options here. It's best to try and avoid buying anything that isn't reusable or returnable in the first instance. Yet having tried the other options then good recycling is better than dumping.

As always, household zero waste starts with your purchasing rather than with your bin. Thus, when considering the recyclability of an item you are looking at on the shop shelves, choose the option that is easily returnable or recyclable back into its original form where possible. Glass is better than plastic; single material plastic better than composite packaging such as 'cardboard' cartons and so on.

With the most recent IPCC[1] report ringing in our ears, we cannot continue to ignore the impact of global weirding. Making disposable items from plastic (from fracked gas, carted around the world in dragon ships, and then burned in incinerators under the guise of recovery) is absolute madness. So remember, recycling is a last resort to be employed with great caution only when all of your other efforts have been exhausted.

Changes to recycling rules

Many recycling rules have changed in the wake of China's decision. The summary of the new rules is 'clean and dry'. One reason for the decision to halt European imports was the ongoing contamination of recyclables with food, nappies and other gunk. As a consequence the list of things generally accepted for recycling has now shrunk to let us get our heads around the 'clean and dry' bit. Perhaps in time the list will broaden out a bit to include other recyclables like polyethylene bags.

In a nutshell many municipal recycling centres now only take hard plastic (excluding polystyrene), beverage cans and food tins, card and paper. As well as being clean and dry, items should be kept loose and separate so that they are easily sorted on a conveyor belt later.

My tendency is to recycle every last scrap of plastic, foil or glossy paper, but it looks as if that's just moving litter from my kitchen to the recycling centre discard bin. For little bits perhaps it's best to simply dump it in the main rubbish bin, and then explore ways to cut it out of the supply chain completely by changing shopping habits and by advocacy at retail, manufacture and policy level. That said, for the small amounts of aluminium that still make it into our house, I still push aluminium foil into drinks cans to keep like metals with like during the recycling centre sorting process.

So should we be recycling at all?

Even if it were possible to recycle all of our plastic waste, the volumes of products and packaging being made mean that there is still too much energy going into the manufacturing and recycling processes overall. The less we have to resort to recycling the better.

We urgently need a complete overhaul of our retail supply infrastructure to ensure that as many of the things we buy are returnable, resalable, or at least compostable. Note however that we cannot simply shift from single-use plastics to single-use bioplastics made from plant materials.

The energy involved in single-use anything is too high for our climate systems, habitats, plants and animals to cope with. Certainly waste-to-energy incineration would be an excellent solution if oil and gas were in endless supply and safe for climate and nature, but they're not. Only by a combination of bans, taxes, deposit schemes and proper reuse infrastructure will we be able to tackle the combined pressures of waste generation, climate breakdown and species extinction.

So should we even consider recycling? Well, in a word, yes. In nature, one species' waste is another species' bonanza. Similarly we will inevitably have materials that will reach the end of their useful life in one industry or application and will need to be either melted down and remoulded, or changed in shape and used as a raw material elsewhere.

That being so, here are some of the common materials accepted at bring centres, and what can be done with them:

Glass needs to be sorted by colour in order to preserve colour differences in the recycled product. Glass recycling is considerably less energy intensive than manufacture from raw silica and other ingredients.

Plate glass, ceramics and Pyrex are not accepted at bottle banks, but there is sometimes a container for them at recycling centres. In my experience, these do not comprise a large volume of our bin, so occasional breakages should not pose a major problem for the landfills of the country. Broken crockery and tiles can be used as fill material for pathway construction, for making decorative and durable mosaic floors or wall finishes, or simply for drainage under pot plants. This isn't recycling back into the original form, but it keeps non-recyclables out of landfill.

Paper, card and newsprint are often collected separately, so keep these separate in your home recycling system to make each visit to the recycling centre quicker and easier. It is even better to get the maximum use out of any given piece of paper, such as using both sides, before recycling. Any cardboard food containers should be washed and dried before folding with the main stack of recyclable card.

Metals usually need to be sorted into either aluminium or ferric metals (i.e. containing iron. A magnet will tell you quickly which is which – it will stick to ferric metals, with iron in them, but not to non-ferric metals). Usually beverage cans are aluminium while food tins are ferric. While metals are eminently recyclable, the industry is a dirty one. Dioxins and other toxins are often generated during the recycling process, particularly if plastics are burned off in the process of recouping the metals. This is particularly so for electronics recycling abroad, often with negligible safety controls.

Textiles and clothing can be sent to second hand shops, charities or a local textile collection depot. Older items can be repurposed for use as cleaning rags, rather than being dumped. See the Bedroom section for more details on clothes and fashion.

Plastics come in a whole variety of colours, grades and types. As a material, plastics are incredibly diverse and useful, but some of the plastic gizmos we produce are remarkably, incredibly unnecessary. I'm sure we'll look back on this time in history with a sense of absolute bewilderment at how squanderous our behaviour was; at how we converted such a rich resource of the stored sunlight in oil and gas reserves into an endless stream of stuff we didn't need at all.

The following list gives a breakdown of the different plastic types that are commonly encountered. However most recycling centres or collection facilities gather all hard plastics together and sort them after collection.

o   PET/PETE No.1 – polyethylene terephthalate; e.g. clear or tinted drink bottles. Readily collected for recycling.
o   HDPE No.2 – high-density polyethylene; e.g. heavy duty liner materials, plastic milk jugs, plastic bags. Hard HDPE is readily collected for recycling. Soft plastics generally not.
o   PVC No.3 – polyvinyl chloride; e.g. PVC windows, sewer pipes, guttering, cling film, blister packs for medicines. Hard PVC is collected for recycling (and recovered for incineration), but is highly toxic when burned and in the environment.
o   LDPE No.4 – low-density polyethylene; e.g. milk container lids, clear food bags. (Light grade, soft plastics are generally not collected for recycling.) Hard LDPE is readily collected for recycling. Soft plastics generally not.
o   PP No.5 – polypropylene; e.g. woven groundcover plastic, coal bags, squeezy ketchup bottles, straws. Hard PP is readily collected for recycling. Soft plastics generally not.
o   PS No.6 – polystyrene; e.g. insulation, meat trays, disposable cups (generally not collected for recycling). Not generally collected for recycling. Toxic when burned and in the environment generally.
o   Other No.7 – Multi-layer plastics and plastics other than the six typical types. Some collected for recycling, but not readily recyclable.

To add awkwardness to the recycling process, different plastic types are often used in the one product, such as HDPE lids and seals on PET bottles.

Remember that if plastics are not adequately washed and sorted then a whole container of materials may be rejected by the collection company and dumped into landfill, so it's important to follow the recycling centre guidelines as far as possible, however inconsistent they can sometimes appear.

Timber (clean) is accepted for chipping in many bring centres. Timber is quite versatile and adaptable for reusing if you are good at DIY. Smaller bits of untreated and unpainted timber can be used as kindling for the fire. What can we do with plywood, chipboard, fibreboard, treated wood and laminates? Due to the glues, resins and waxes used in these composite boards, these should not be burned or composted. Avoiding them is not necessarily the answer either, since they mop up a whole lot of waste wood bits from the timber industry. This means we get more boards per forest than we would otherwise; between sawn planks and composite-wood products such as chipboard. Increasingly non-toxic binders and resins are being developed to make these products safer.

For existing timber, the most appropriate solution would be for bring-sites to sort wood into natural wood suitable for chipping for landscaping purposes or fuel, and composite wood suitable for recycling into new composite board. This is already the case in Germany and the UK, facilitating the recycling of composite board products. The National Community Wood Recycling Project is an example of local timber recycling in action. They have collection centres for waste wood and for reclaimed timber sales throughout the UK[32] and could easily be replicated anywhere in the world.

Mixed material products and other non-recyclable: some products and packaging contain a combination of materials that are not easily separated. The most common are Tetra Pak and other foil or plastic backed 'cardboard' cartons. These contain cardboard with thin layers of plastic and sometimes aluminium. Tetra Pak cartons and other card-metal-plastic composites are currently collected with cardboard. However, the metal and plastic in them makes them impossible to fully recycle because they cannot to separated out into their constituent materials. Despite being labelled recyclable, composite cartons can't be melted down to make more composite cartons. They can only be made into materials of a lower grade, at best. In some specialist recycling plants the carton materials can be mushed up in a wet mix and the cardboard then recouped from the aluminium and plastic, which is landfilled. Otherwise the whole carton is removed as a contaminant in the paper stream at the point of recycling, and dumped. So much for bring recyclable! Avoid these on the shop shelves altogether, and opt for returnable, reusable or fully recyclable containers instead.

Products like waste electronic equipment also fall into the category of mixed material products. The recycling process can be a particularly toxic and hazardous one, particularly when the plastic element of the waste is simply burned off to reclaim the metals. Keep and use your electronic equipment for as long as you can, and consider second hand equipment when making any new purchase to minimise the volume of waste electronic equipment at large in the world.

Setting up the structures for recycling

Different recycling centres will require different levels of sorting. For maximum ease at the centre, you may wish to have a different container at home for:

○ paper
○ card
○ hard plastic
○ aluminium cans
○ food tins
○ glass
○ a general box for other items such as batteries, ferric metals, CDs, polystyrene and the like.

We have a box in the kitchen for general recyclables, which is sorted later into the different categories in the shed. If you are keeping clean uncoloured non-glossy paper to start the fire, this can be stored in a wastepaper basket near the source of generation – as long as everyone knows that the basket is not for plastics or glossies or receipts (heat printed receipts are coated in the toxic chemical BPA and best kept out of both the fireplace and the recycling centre).

Very soon after starting your reduction measures in earnest, you will discover which categories need the most space, which containers can be doubled up, and which are redundant. Change your system as often as you need to, and be imaginative about the set-up. Don't start with enormous brand new plastic boxes for each material if one old cardboard box will do for glass and metals and another box for everything else. When you find out what your needs are then you can refine the set-up as you go along. Probably best not to rush out and consume a barrel load more resources (and future waste products) for your waste minimisation endeavour.

Sourcing recycled products

To help create demand for recycled products it makes sense to buy them rather than buying virgin material products. Not all recycled products are the same. Like shopping for good food, always read the label and get to know your supplier. For paper products, look for 100% post consumer recycled paper; unbleached or oxygen bleached so that you avoid the ecological damage done by the chlorine bleaching process.

In the long term, recycling might be a bit of a dead end detour on the road towards zero waste. However we're in an interval between the blatant waste of anything and everything at present, and a good returnables infrastructure that negates the need for most recycling. While we're in this gap the best we can do is support those companies that are offering good recycled products and keep pushing for change through advocacy.

Step 5 – Give Feedback, Push for Change

We can apply zero waste measures in our own homes, but we cannot snap our fingers and instantly magic up the society that we want. While we work we have to be patient as well. The big changes happen when government policies and industry supply chains change for the better. We can play an active part by encouraging governments and companies to do the right thing and introduce the structures that would help us have a zero waste society.

Despite the seeming challenges, positive change is possible. In fact, it is rolling along steadily with each generation. Push for change in policy at local, national and international level. Talk with traders and suppliers about zero waste and encourage them to supply you with the goods and services you need without the extraneous plastic thrown in.

Give feedback to companies whose packaging continues to make its way into your bin. Tell them you want something that doesn't cost the Earth. They're offering you a service and many will take on board the concerns of their customers.

Look for petition sites that push government policy and sign the ones that you think are important. With so many paid lobbyists working at EU level it's vital to counter the economic priorities with good protections for the biosphere itself. Without the living systems in the world, we can have no economic activity anyway, so we may as well push to have that acknowledged in our legislation and government policies.

There's a section on Advocacy at the end of the book that provides some extra background, hints and ideas in this area.

Zero Waste in Action

There's nothing like just jumping in and getting started. Start by checking what is in your bin. For this exercise include both your main rubbish bin and the bin for clean recyclables. When it comes to plastic waste and resource consumption, both bins have an impact on the world. Recycling isn't a get-out-of-jail card.

On a sheet of (scrap) paper, write down the items or categories that are present. Sort by volume, starting with 1 for the category with the largest volume and working down from there. Here's what a typical list might look like:

Rubbish bin			Recycling bin	
Item/category	Bulk ranking		Item/category	Bulk ranking
Clear plastic wrapping	1		Milk bottles	1
Nappies	2		'Cardboard' cartons	2
Pet food bags/sachets	3		Newspapers	3
Cereal bags	4		Plastic fruit boxes	4
Jam jar lids	5		Jam jars	5
Yoghurt tops	6		Shampoo bottle	6
Medicine blister packs	7		Batteries	7

Next take the item that is most common in each of the two lists and start to explore ways to reduce or eliminate those from your waste stream.

In the example above, the main item in the rubbish bin is clear plastic bags – the kind of wrapping that comes around bread, fruit or veg, whether from the supermarket or the farmers' market. The main item in the recycling bin is plastic milk bottles. So how do we go about eliminating or reducing item 1 on each list?

Remember the list of actions in the previous chapter – starting with mindfulness while shopping. In this instance, can we refuse to take in the items in question? For the polyethylene bags, which come from so many different sources around so many different groceries, it's tricky to refuse them all. Certainly we can make a good stab at it though. The plastic bag levy means that reusing shopping bags has rapidly become the norm. But small plastic bags still come into the house via purchases of bread, loose fruit or veg, meat, fish and so on.

We can examine each of these areas in turn and take the first step of refusing them where possible. We can give market traders their bags back if they're willing to take them and reuse them. Supermarkets may take them back, but won't be able to reuse them. You can leave them behind in the supermarket bin by all means, but actually getting them back into use again is where the ecological savings are to be made. Perhaps give the supermarket a miss if you can't get what you need without getting plastic thrown in. Our experience has been that our diet changed for the better over time as we refused more food in plastic; making substitutions for more local, fresh and organic produce.

Fish and meat can be brought home in a plastic box or large jar rather than using a new bag each time. I find that fishmongers and butchers sometimes weigh the box first and then calculate the weight of the product separately so you don't get charged for both the box and the meat or fish, but not always. For me, the priority is to reduce the plastic waste rather than to have the price deducted, so given a choice, I'll ignore the charge, but will be quicker to return to those shops that account for the weight of my box.

Generally when looking at household waste, the kitchen is where most gold will be mined and the most changes made. However, other areas should also be considered. Does your everyday soap come in a plastic wrap? How about the polythene bag that the Sunday supplement comes in, hidden in amongst the main newspaper? Have a look at the other areas where polyethylene bags make their way into the house and see if you can refuse those too. You may not manage to cut out all of them, so realistically you'll probably reduce this category rather than eliminate it completely.

After refuse and reduce, comes **research**. Ask around, go online, join a local group to help you as you find out ways to get closer to zero waste. Email me if you're stuck and looking for ideas.

Next up is **reuse**. Although you might not immediately consider it, polyethylene bags can be washed, dried and reused many times before finding their way to the bin. They're useful things – which is one reason why there are so many of them. By having a stock of washed dry bags to hand, you can save yourself buying a roll of freezer bags or avoid having to get a new bag when caught out shopping without a box. They're tricky to dry, but we just up-end them on the draining rack with the crockery and then hang them up on a decorative clothes peg glued to the kitchen dresser to dry after they've drained a little.

Repairing a plastic bag with a hole in it probably won't find too many enthusiastic adherents – but for things like bicycles, ironing boards, shoes, crockery or what have you, it can be a very effective way to keep them out of the bin and prevent or delay the purchase of a new replacement.

Composting won't apply to plastics, so we'll skip it for this example.

Recycling is towards the tail end of our list of options. Soft plastics like polyethylene bags aren't one of the resources collected in our local area, so this option may or may not be available to you. Even if your locality does provide plastic collection for bags, now that China has stopped importing EU plastic for recycling, there isn't necessarily the infrastructure here in Europe to recycle it all back into a useable resource again. Other Asian countries now receiving European waste plastic for recycling are not necessarily as well equipped or organised as China was, so the potential for leakage of European and US plastic waste into Asia's rivers and the world's oceans is higher than ever before.

Recouping the energy from clean wood and paper in a local waste-to-energy incinerator or for starting the fire can be a good way to recoup some value from waste materials. However this is only safe if toxic plastics (such as PVC and polystyrene) are specifically excluded, since burning these plastics has a high potential to release very harmful pollutants. Also, household burning of waste is illegal in Ireland and causing a nuisance by doing so is prohibited in the UK.

Finally, please **remember** that this is a process. Life isn't perfect. We won't manage to have zero impact or produce zero waste within this society of ours just yet. Do your best, store the rest and let it go to the local landfill site or incinerator when the container is full. You may not be able to get rid of all of your plastic waste or even all your polyethylene bags, but you might achieve a reduction of 90% or so. Pop the rest into the largest of the bags you are throwing out and next time you are passing the local dump, drop in and pay your few euros to let the council mind it for the next several thousand years. The longer you get between trips to the dump the better. Don't beat yourself up for any shortfall. Better to have fun with the process than to break yourself pushing the waste issue up hill.

So, in a nutshell, we've looked at the source of the waste and then explored ways to eliminate it from the bin, or at lease reduce it. Do this for the top items in both bins. Chapter 6 has lots of ideas as we explore the house room by room.

Try this for a week or two and see how it goes. See how your shopping habits change. Watch for opportunities to make the process sociable rather than a nuisance. The former is certainly more fun than the latter, and it's usually just a frame of mind that makes the difference.

If this looks like a daunting task, then another approach is to take the lists of wastes generated and see what's easy to remove rather than the item with the greatest volume. It really doesn't matter how you start. The trick is to jump right in and begin.

Don't worry about being too consistent in your approach. Recognise that you will be happy to reduce in some areas and not in others. Maybe you will be quite happy to give up the daily paper as a way to reduce your waste paper, but really like to read novels by the score. Do what you are comfortable with and be imaginative about your approach. (Maybe the library would provide the novels and you could reduce both sources of paper consumption!)

Since Elinor and I have lived together, we've been limiting our waste using a combination of these measures. That doesn't mean we do it perfectly by any means, but we've certainly put thought into the process. I've seen a steady progression from initially trying to buy products in recyclable packaging, to recognising that recycling is very limited in its environmental benefits and moving towards reducing packaging completely where we can.

We started by cutting down on plastic wrapping around dry goods, plastic food trays and by shopping more in the local markets. We'd never really been PET water bottle people, but for airport travel that was still an issue, so we figured out that if you carry an empty long-life bottle through security you can usually fill it again on the other side. Next we saw the need to minimise the need for new battery operated gizmos and electrical appliances to cut down on hazardous and electrical wastes – even for rechargeable batteries, which don't last forever. We have migrated away from buying certain products so that we avoid the packaging or product itself, and try to get products closer to home if possible to reduce the waste associated with longer food miles.

You'll find different products and containers in your own home, so apply the general principles and see how you get on. It's a process. Have a go and have fun with it.

Should I Bin My Bin?

We've not had a bin collection service for our household for the past 20 years or so. It hasn't been a particularly difficult decision. We have had nothing close to zero waste, but the volumes that we produce probably fill a standard wheel bin every six months, plus about 2-3 times that in recyclables. Paying for a full annual collection service (>300euros/yr in my area of Ireland) for about 10% of the waste volume that most households produce simply never made sense to us.

If the above measures look manageable, then the question of whether to keep your bin collection service is a bit of a moot point. It simply ceases to be necessary. However it's easier in some places than others. If you live in an apartment without access to community composting, then you'll find it trickier to deal with food waste than if you are an avid gardener. But there are ways around most challenges. Befriend a gardener for example.

If you don't have temporary storage space for recyclables then convert your existing bin into a recycling store and go to the local bring centre a little more regularly. Another option is to share your bin service with a neighbour who has the same aims.

If you live in an area with collection fees, try moving towards zero waste between now and the next bin collection bill. See how you get on and set your sights on a lower impact future and lower waste bills.

If your waste collection service is provided free of charge then by all means keep your bin, but note the dates that you put it out so that you can measure how little waste you have after making the changes suggested in this book.

Chapter 6

Zero Waste, Room by Room

The ideas in the preceding pages offer a general overview of the process of moving towards zero waste. In this section we explore reduction options more specific to different areas of life.

What follows is only a list of ideas. Keep your eyes, ears and mind open, and use your imagination liberally. The process can be enjoyable. If working towards zero waste and a healthier world feels like a game then you'll be an inspiration to others, and have more fun along the way.

The Kitchen – Food and Food Preparation

The kitchen is the hub of the home. This is where we provide ourselves with sustenance, nurture, variety and flavour. I grew up in a family where the kitchen was literally and metaphorically the centre of the house.

For all I cared and knew, the world revolved around the kitchen. It is also the centre of the action when it comes to waste generation. If you look in your bin you'll probably be able to attribute 90% of the stuff in there to cooking. Tins, plastic trays, milk cartons, PVC film, polythene bags, styrofoam trays – and that's just the packaging. There's also leftovers, fruit and vegetable trimmings, spoilage, grill pan fat and other food waste that needs to be considered.

Here are some ideas to get you started towards zero waste in the kitchen:

Shopping: where it all starts

Only put in your shopping basket what you want to eat. You can't eat the packaging! Choose loose fruit and veg. Opt for raw ingredients rather than heavily packaged ready-made meals. Next best is to select packaging materials that are easily recyclable or have a low embodied energy and low food miles. Single material packaging is more readily recycled – or packaging that can be readily separated (such as Glenisk or Yeo Valley yoghurt cartons) rather than composites (such as Tetra Pak containers).

When shopping, use cotton, jute or hemp bags rather than plastic or synthetic ones – even the long life versions. That way you can compost the bag when it finally meets the end of its life. While avoiding plastic is the primary goal, even paper bags have a significant carbon and resource footprint. Carry a stock of your own small washed bags to avoid taking in new ones. If you're good at sewing, it's easy to run up a handful of drawstring bags from old fabric that may otherwise end up in the bin. Bring along a box or jar for fish and meat.

Soft wrapping

Soft polyethylene plastic wrapping is the waste that seems to be most problematic for people moving towards zero waste. If your usual shop only sells fruit and veg in plastic then talk with them about it and explain politely and encouragingly why you want to see a change. If they continue with plastic then try other shops or a local farmers' market in your area. You may also wish to grow some of your own fruit and veg.

Liquid products such as milk and yoghurt are trickier to buy plastic-free. Some areas have glass bottle deliveries from a mainstream dairy or from local farmers, so do some research and ask around in your local health food shop. They're usually a mine of information. Alternatively buy dry loose nuts or oats for homemade veggie milks. Yoghurt is sometimes sold in glass rather than plastic. Remember that glass still has a high carbon footprint unless it is reusable, so ideally return the containers to the supplier if you can.

If you think that cheese only comes in plastic, then you have a whole new culinary adventure awaiting! Farmers' markets, specialist suppliers and good supermarkets all have excellent cheese counters now, so check out the variety of fare on offer and bring your own container. Apart from being unnecessary plastic, the food soaks up chemicals from the cling wrap.[33] If you know a cheesemaker, try ordering a whole round. Cheese freezes well in parchment paper so you can eat it at your leisure, and the paper can be washed multiple times for even less waste.

For bread, try a local bakery if your usual shop only sells in plastic wrapping. For dry ingredients, look out for paper packaging, or find a local supplier of loose pulses, nuts and seeds and bring your own container.

Other packaging: cans, tins, plastic trays and containers

We need to recycle less, not more. Although glass, metal and hard plastic can be recycled, they all take their toll on the natural world. Single-use packaging of any sort has a higher carbon footprint than a reusable container. Cans and tins are a case in point. To cut out metal waste you can select dried beans and pulses rather than canned; fresh fish rather than tinned. To cut back on hard plastics, avoid foods sitting on a tray. Packaged food is often more highly processed, which isn't all that good for you anyway, so your health may improve as your environmental choices change.

Try cutting packaging out completely for a week as an experiment. This is easier than it might first sound. You can bring a lunch box to the fishmonger or butcher to avoid the plastic bag. You can make your own granola using ingredients bought in bulk, in paper, or from a scoop shop to avoid the plastic wrapped breakfast cereal. Porridge oats can be bough in a plain paper bag and are local and traditional for these islands. The ideas are endless, so have a look at your diet and the packaging it comes in and find a way to change one or the other, or both, to wriggle out of the plastic noose.

If cutting out all packaging entirely sounds like a stretch too far, try permanently removing one packaged item per week. Finding suitable alternatives can become more of a game and may thus lead to more lasting changes in shopping habits.

A final point on packaging; find tea bags without plastic in them. Most tea bags have plastic included in the paper mix to keep them together better.

Whatever the quality of your tea, this means that when you put them in the compost, worms can't burrow in and make use of the tea leaves. Find a supplier of plastic-free tea bags, or switch to loose tea leaves.

Recycle what's left

Even if you cut out plastic entirely, you'll likely have some packaging remaining at the end of the month. The most easily recyclable materials are cardboard and paper; glass and metal. I've found it really tempting to recycle every last scrap of aluminium foil and plastic ring, but in reality small bits will fall through the sorting machines and get discarded as waste for landfill or incineration. Likewise polyethylene bags tend to get fouled up in conveyor belts and machinery, costing time and money for the recycling firms. Cut down or cut out these at source if you can.

Unwashed food containers or contaminants such as nappies have led to whole batches of otherwise recyclable material being rejected and dumped. Wash and dry all hard plastics, glass, cans and tins for recycling. Separate loose paper from food tins and send each to their separate recycling streams.

While recycling has its own ecological footprint, its a useful way to cut out the bin habit until such time as we develop proper return infrastructure in our society.

Avoiding food waste

About a third of all food produced or raised never makes it to your plate. Much of this is dumped before getting to the supermarket shelves because it doesn't meet the required 'quality standards'. This wastes energy, water, land, seeds, labour and money on a huge scale. About 8% of all global greenhouse gas emissions are attributable to food waste, making it the third single largest contributor to climate breakdown.[34]

Most of this waste takes place before ever a carrot or apple reaches your kitchen, but where its cousins and siblings were junked because they weren't pretty enough. Think about it. By simply reducing global food waste and buying wonky veg along with the straight, we could see massive reductions in carbon emissions. It is also one of many measures that could free up land for rewilding projects, large and small – from new wilderness areas to wider hedges on our farms. Along with avoiding food waste we can avoid wasting farming space growing cereals for cattle feed, and pasture them instead. This is a healthier option for cattle and with holistic management can also be a valuable carbon sink we well as preventing soil erosion.

Speaking of cattle feed, cows do best on pasture – preferably a good mixed sward with a wide variety of plants. Isabel Tree's descriptions of the cattle on her farm wilding project speaks of the wisdom that the herd has in selecting food that is good for it across the wide range of rewilded land available to them.[35] From a waste perspective, when you feed cattle anything other than pasture you quickly lose control of your food chain and what goes into it. Feeding meat and bonemeal to cattle has been banned since the BSE debacle, but that space has been filled with other stuff I'd rather not be ingesting either. Genetically engineered soy, maize and other crops[36] for one, but also the leftovers from supermarket food waste – complete with the wrapping chopped up into the feed pellets![37] A good reason to find a local regenerative beef and dairy farmer if ever there was one.

In our own homes we can address global waste by making sure that the food we buy gets used. Best-before labels are indicative and are often quite flexible. Use your nose and sense of taste rather than dumping stuff that is perfectly good, but out of date. Don't overdo it though and eat spoiled or rotten food. Having made this mistake unwittingly myself I can't say I recommend it!

It's easy to take simple steps to reduce food waste in the kitchen. Make broth from leftover bones; use a spatula to get the last bit out of jars and saucepans; find new recipes for cooking leftovers; learn to read best-before dates with a pinch of salt and use your senses to assess whether food is good to eat or not. Rinse nearly empty jam or honey jars into sauces that need sweetening.

On a wider level, food waste is an issue for advocacy and of carefully selecting where you spend your money. By shopping at local markets or small grocers shops you'll be supporting those traders who really cut back on food waste at the wholesale and retail stage. If markets and small grocers aren't readily available in your area, then write to the CEO of your regular supermarket and ask them what they are doing to minimise food waste across the whole supply chain from field to fork. It won't be the first letter or email they get, but it may be the one to make the difference.

Food storage

Proper storage prevents spoilage and wastage of food. It should be unnecessary to buy freezer bags, since enough suitable bags usually come in with the shopping around things like cornflakes, bread, dried beans and the like. Large yoghurt cartons and 'cardboard' cartons can be repurposed to make freezer boxes for soups, stews and stock and can be easily torn away for reheating food, or used again if the food is left to thaw first. Note that both of these will inevitably end up in the bin (recycling or landfill) anyway, so they are best avoided at source rather than even repurposed.

If you preserve your own food you can reuse old jars and bottles as well as cutting down on cost and waste. Where food is stored on shelves, ensure that it is reasonably cool and out of direct sunlight to avoid spoiling. We've found that large peanut butter jars work a treat for storing dry ingredients, but when we ran out of shelf space for new jars we bought a better blender (to replace the old one that was on its last legs). Now we can make nut butter ourselves from scoop-shop ingredients so that we can avoid even the jars. Actually, for some reason we can't get loose organic peanuts – even from our scoop shop, so these come in a plastic bag. It's a work in progress.

Refrigerant chemicals (used in fridges, freezers and air conditioners) are the largest single source of global greenhouse gas.[17] Cool and freeze your food by all means, but the longer you avoid purchasing a new fridge or freezer the more you help in global efforts to reduce the effects of climate breakdown. Consider storing veg in a north-facing shed or store room to keep it cool, which helps to keep your fridge space requirements lower. At end of its life, be careful to ensure that your fridge or freezer is sent for proper recycling and recouping of the refrigerant gasses.

Minimalism has its advantages when it comes to storage. The less you store, the fewer storage containers and less storage space you need. If we all had smaller houses and simplified our lives, we'd all reduce our ecological footprint in one easy step!

Food preparation: vegetable peelings and trimmings, meat trimmings, bread and pastry crumbs

The first step here is to be judicious about what you select for the pot and what for the compost bin. Clearly the greater proportion of food you can put in the pot, the lower the waste and ecological footprint of your groceries.

If you buy from local growers who take pride in the quality of their produce they wouldn't dream of throwing away perfectly good veg just because it is a little knobbly. If you grow your own you'll see a share of unusual shapes. As a result, market or homegrown veg often needs a bit more prep in the kitchen, so the compost bin will be your new friend. Compost vegetable trimmings, fruit cores and stones, as well as any leftover bits that can't be used for the next meal. Even kitchen sink drain sieve bits can happily go into the compost.

Meat trimmings can be fed to pets. They thrive on a varied diet just like we do. Bread and pastry crumbs can go on the bird table to encourage feathered friends to stay within sight of the kitchen sink for company during household chores. It is sometimes recommended to clear the bird table at night (into a good compost system) to avoid encouraging rodents, but a good tall table will also work. If you want to go the extra mile, hens are a great way to convert scraps into protein and can be kept in even a modest sized garden.

Cooking: cooked food, bones, oil, fat and leftovers

Bones of all sorts and clean vegetable peelings and trimmings can be made into stock, or frozen until there is a sufficient quantity to do so. For best results roast the bones in the oven and then decant into a large saucepan and cover with water. Wash the lovely caramelised juices off the pan by heating over a gas flame and stirring with a wooden spoon to lift all that flavour off for your stock. Add some peppercorns, a carrot (or the tops and tails of the carrots if you prefer) and a quartered onion for flavour, then a handful of garden herbs. Skim off the scum that forms as the stock starts to boil, to remove impurities. Boil for anywhere from 4-24 hours and strain. Bones and other sieved bits can all be thrown into the compost with everything else. Cooked food in modest amounts isn't a problem for composting if the system you use keeps out rodents.

When you cool meat stock you'll have a layer of good healthy cooking fat that will store in the fridge. If you have waste fat or oils after frying or grilling, for example, don't put them down the drain where they can quickly accumulate and lead to blockages. If you're not going to eat it you can use grill pan fat in newspaper to light the fire. Used cooking oil can also be soaked in paper as lighters; returned to some council bring site; or find a friend with a converted diesel engine that is suitable for using vegetable oil. If you have too much meat fat you can make bird feeders by pouring melted leftover fat into egg cartons or coconut shells full of nuts and seeds. Hang upside down from the bird table when solid, to keep the birds going through the winter.

To avoid leftovers, don't cook too much to eat at a given meal.

Keep any leftovers still in the pan or dish, rather than dumping – either freeze or refrigerate, and reuse at a later meal. Use a spatula for getting pans really clean, so you get to eat all that you cook rather than washing it down the sink and leading to extra water pollution. We've been moving from plastic to wooden implements in the kitchen, but I have yet to find a comfortable balance between getting the very last out of a saucepan with a good plastic spatula and converting entirely to wood. One solution is to take the saucepan with the sauciest dish (tricky to wash) and start the next day's soup in it.

A zero plastic kitchen?

Talk of plastic spatulas brings us to another question: can we have a completely plastic-free kitchen? What would we do for light fittings, cookers, the fridge, blenders and utensils? Even paints, floor coverings, taps and curtains have their share of plastics or additive... Alternatives can be found for many things (everything in fact, given that plastic is a pretty new invention in the greater scheme of it), but one easy thing to consider is that if you already have something in use, then the resources have already been invested in making and transporting it to your home. If everyone were to extend the life of their equipment, that would lead to a considerable reduction in new things being made and sold in the world. That's a considerable ecological saving.

When we removed the plastic implements and utensils from our kitchen recently, we hardly missed them. Like many other people in our culture, we had many duplicates, and so we kept the wooden or metal ones and let go most of the rest.

While we may not achieve a zero plastic kitchen, a good declutter is certainly welcome. If we let go the items that are surplus to requirements rather than binning or hoarding them, they can displace a brand new purchase by somebody else. This helps to keep down the energy and resource inputs needed to manufacture new products.

Diet: what goes in your shopping basket goes into you

How does diet fit into a waste minimisation plan? Generally a reduction in packaging coincides with a reduction in highly-processed foods, which are also generally high in salt, sugar, trans fats and additives. An organic, wholefood diet is usually better for the natural world and for your health, and can often have less packaging to boot. A frequent complaint is that it is more expensive, but that's balanced by savings on bulk buying and on unprocessed foods.

If you go the extra mile on the zero waste journey and grow your own food (using your own compost) then your diet will benefit not only from good wholesome food, but from unparalleled freshness. You'll also benefit from the sunshine, fresh air and movement that go with growing.

Drink water. Just tap water. By all means advocate to keep it free of fluoride and breakdown products of chlorination, or filter it if you need to, but it is much better for the world than the growing tide of plastic bottles.

Health

A number of years back in the lead-up to our national elections, a group of candidates was being asked whether our healthcare system should be better funded. What is a candidate to say? One after the other they gave their reasons why, absolutely yes, our public health system should be in receipt of more money. Until the last candidate spoke. Fergal Smith, a local surfer, farmer and Green candidate at the time, bucked the trend. To paraphrase, he said something like "Why are people getting sick? We don't need to put more money into hospitals and medicines, we need to keep people healthy. Put money into getting healthy food on their plates so they don't end up sick in the first place". He's right. Our medical model of health simply isn't working. Look at the epidemic of cancer, diabetes and other diet and lifestyle related illnesses. People need good food that's been grown without chemicals so that it can fully absorb the nutrients and minerals that the soil has to offer. We need exercise and fresh air and sunshine. We need lifestyles that support health, rather than medicine systems that manage illness.

I debated putting a medicine in the bathroom section, where many people keep their medicine cabinet. In the end though I decided I'd be better off sticking health in the kitchen where it belongs. There are so many different dietary guidelines out there that it can be tricky to know where to start. However some simple guidelines might be helpful. Notice that the fresher and healthier a food, the lower the waste seems to be as well.

O   Avoid food with poison on it. That might sound pretty self evident, but conventionally grown crops and livestock are routinely dosed in chemicals to kill off other plants, insects, bacteria and/or fungi.

Needless to say, chemicals designed to kill living organisms won't be all that good for your body. This applies to artificial sweeteners in 'diet' food and drinks. They're not designed to kill things, but can be neurotoxic and carcinogenic nonetheless.[38]

○ Eat real food. Isn't all food real? If the scarily durable fast food meal put on display in an Icelandic museum from 2009 to 2015[39] is anything to go by, no. Real food simply means food that has been grown or reared naturally and then had the minimum of processing between farm and fork.

○ Eat food that is as fresh as possible. Processing and packaging can certainly extend the shelf-life of food, but they don't tend to preserve the vitality your body needs to stay healthy. Bear in mind that generally the greater the diversity of colours on your plate, the greater the mix of vitamins and minerals present.[40]

○ Eat healthy fats such as organic extra virgin olive oil, organic grass-fed butter, organic meat fat or organic coconut oil. Olive oil is one of the most adulterated foods on the planet,[41] so look for a supplier who knows the grower if you can. Fats are enjoying a renaissance after a couple of generations of misinformation about saturated fats.[42] Nina Plank's Real Food[43] offers good solid scientific guidance on this subject.

○ Choose your mind-altering substances with care. These include sugar, caffeine, alcohol and a host of other legal and illicit drugs. If you need a kick to get going (I'm going to run the risk of sounding like an agony aunt here) have a good look at the circumstances in your life and see if you need a new job, a change of situation, or just a good rest rather than continuing to prop yourself up with substances.

○ Enjoy offal. Good chemical-free organ meats are amongst the most nutrient dense foods you can eat.[44]

○ Enjoy your food and treat it as a sacred gift. The pleasure with which we eat has a direct bearing on the health of our body and soul. We can learn a lot from French mealtimes; leisurely, sociable and enjoyable. Revel in the flavours, textures, colours and aromas of good food.

○ Fresh air, sunshine,[45] exercise and good company are all important ingredients in the mix too.

When dealing with illness (as opposed to maintaining health) natural medicines and tried and tested home remedies will have a smaller waste footprint than conventional medicines. Many effective remedies exist already in your kitchen cupboards. As with many areas of modern life, the conventional approaches are not always the kindest to either you or the world around you, so check out the many traditional and complimentary practitioners in your area and see what works best for you and your family.

If you already have conventional medicines in your home that are out of date, take them to your local pharmacy for safe disposal.

Wastewater and energy issues in the kitchen

Think beyond the bin and consider energy use and wastewater, particularly in the kitchen. Allow food to cool before refrigerating or freezing, to avoid wasting energy. Steer clear of garbage grinders because they increase the wastewater pollution leaving the sink.

Water conservation is another way to reduce your waste, although it's more difficult to see directly. Take care not to leave taps running, and to fix any leaky ones that drip. A surprising volume of water can be wasted just one drip at a time. Use particularly dirty wash-up basinfuls to water a hedge rather than adding to the drains; or particularly clean basins to water your patio herbs.

How you cook also has a direct bearing on energy use. For example proper soaking of rice and pulses before cooking cuts down cooking time. Straw or haybox cookers can be used to slow cook pot stews for dinner after just an initial boil in the morning. We have an old timber wine box kitted out with some old wadding as a mini haybox cooker for pot stews. We pop in the boiling pot and cover with a tea towel and a purpose trimmed folded blanket when we need to have the next meal ready and waiting.

Social and ethical issues beyond the kitchen

Although not a waste issue exactly, consider the wider implications of your shopping choices. Imported items like chocolate, tea and coffee are often produced by growers who are paid a pittance and who often have to put up with unhealthy work conditions. Look for Fairtrade alternatives to guarantee that your money is supporting ethical producers. Many towns and cities are now Fairtrade towns, so they are easier to find than ever before. When shopping use your LOAF (Local, Organic, Animal-friendly, Fairtrade) as a reminder.

The Kitchen – Washing and Cleaning

Moving on from food and cooking there is another activity that is in the domain of the kitchen in most houses, washing and cleaning.

Cleaning products

Cleaning products such as soap, washing-up liquid, dishwasher detergent, scouring powder, window cleaner etc. can all be substituted with safer, less environmentally impacting and less waste generating products. Bread soda (bicarbonate of soda) makes a very good scouring powder. Environmentally-friendly soap, washing-up liquid and dishwasher detergent are available in good health food shops; often as refills to avoid new plastic. A mixture of hot water and vinegar makes a good window cleaner and general degreaser. A couple of drops of tea tree oil can be used as an antibacterial, antifungal ingredient in wash water where necessary.

Many of the standard cleaning products that we use around the home are, in fact, quite harmful for us. Yet they can enter the body via the skin when we are washing and cleaning, and can remain on our eating utensils to be ingested. By using vinegar, tea tree oil, bread soda (bicarbonate of soda), soap and hot water in various combinations we can get the house spic and span without using anything harmful in the process. These are all available locally between your grocer and health food shop.

To avoid paper towels you can use a washable cloth, ideally made from old cotton clothing that isn't good enough for the charity/thrift shop. Old t-shirts are excellent. Use cotton because artificial fibre clothing will shed microplastics into the drain every time you wring it out.

In terms of cleaning utensils there is a growing range of ways to avoid plastic in the kitchen. Stainless steel pot scourers are more durable than the plastic equivalent. Coconut fibre scrubbers make a more eco-friendly version. Loofahs are the fibrous insides of a type of marrow (so there's no reason why you couldn't grow them, although I've yet to try). Eco-friendly scrubbers are becoming much more widely available. Many health food shops have a whole array of zero waste implements for cleaning and there are more online. Timber scrubbing brushes are also becoming more readily available. Some have natural bristles but most still have plastic, which will still end up in the bin.

Packaging

Unlike some food items, cleaning products are generally not grossly over-packaged, but try to avoid packaging where possible. Some health food shops offer a refill service for detergents and wash-up liquid. Next best is to find packaging materials that you can readily recycle locally. Always look for single material packaging or packaging that can be easily separated for recycling. Concentrated products can cut down on packaging, but be sure to use only the right amount, or less, to avoid putting excess down the drain inadvertently.

Wastewater considerations

Every cleaner or chemical that you use in your sink or on surfaces around the house ultimately flows down the drain. Whether you have a septic tank or a sewer connection, these chemicals ultimately end up in the groundwater, your local river, or the sea.

Cleaning chemicals can have toxic and harmful ingredients such as heavy metals, which do not vanish during sewage treatment, but remain in the effluent or the sludge and then accumulate in the natural world. If you use food products such as vinegar and bread soda (bicarbonate of soda) to clean, this has less impact on your groundwater than chemical cleaners. If you must use whatever it is that you 'can't live without', try to minimise it. Just a dash rather than a generous squirt will at least halve your chemical usage, halve your bills and halve your environmental impact. That said, some toxins in day-to-day products can be detrimental to aquatic wildlife at tiny concentrations, so if you wouldn't eat it or grow your food in it, it's probably best not to wash it down the drain.

Keeping house

When washing windows, vacuuming floors, dusting behind ornaments etc. the same low-waste measures can all apply. Windows can be made sparkling with some vinegar, or a soap/vinegar mix in an old spray bottle, and use old newspapers to bring up a shine. Bagless vacuum cleaners allow you to put your dust bunnies into the compost, but even the bags can be emptied and reused with a little more work. The more you use a sweeping brush and dustpan the longer you'll get from your vacuum cleaner – or if you steer clear of carpets and rugs you might be able to adopt the low-tech option permanently.

The Bathroom

The last 10 years or so has seen a great increase in awareness around zero waste. The bathroom is one room where this change is quite visible, evidenced by the availability of cosmetics, detergents, soaps and shampoos that aspire to take their place in the zero waste household.

Hand in hand with finding substitutes for the usual products you can reduce waste be making your own. Also avoid using more of a product than is necessary. If you cut back your use of a product by half, that's also a 50% saving on all packaging and transport needed. Try to source products that have returnable packaging and/or products that have only natural materials so that you minimise embodied waste, i.e. waste created during the manufacturing process.

Stuff to keep the bathroom clean

As in the kitchen, bread soda (bicarbonate of soda) makes an excellent scouring powder and vinegar a universal degreaser. Tea tree oil makes a good antibacterial agent and unlike chlorine bleach, it won't kill off essential septic tank bacteria. Natural citrus-based toilet cleaners can be used instead of conventional products.

Our home is in a high limestone area in Co. Clare, which makes for mineral rich drinking water, but plenty limescale in the toilet bowl. Given that I've got an aversion to putting chlorine bleach in the toilet, the bowl does tend to scale up a bit. I tried a strong acid cleaner once, which worked as it said on the bottle, but I wasn't sure what the wider implications were during manufacturing and use, so wasn't entirely comfortable with using it. Somebody suggested hot vinegar. I heated up enough to fill the toilet bowl, removed the water and quickly replaced with a saucepanful (so that the sewers weren't open for long) and left it overnight. It worked a treat. Not necessary quite as effective as the acid cleaner with its host of warnings, but with a little encouragement from a strong twig (locally grown and zero waste!) it brought the bowl up just like new.

Stuff to keep you clean

Using soap bars rather than liquid soaps will cut out the unnecessary dispenser bottle. Unwrapped, natural soaps can be used to further cut down packaging and embodied waste. Some suppliers now package their parcels in compostable materials. Good health food shops or online suppliers now offer shampoo bars as well. Alternatively (if you find that these don't give the results you're used to) ask your health food shop to refill your shampoo bottles. Our local supplier here in Co. Clare will refill for less than of the cost of the same product in a new bottle.

Dental hygiene is another area where the market has shifted in the past

10 years. It's now possible to buy bamboo tooth brushes and silk dental tape in cardboard or metal containers. Bamboo tooth brushes often have plastic bristles though, so pig hair is the pre-plastic alternative.[46] Needless to say, steer clear of toothpaste with microplastics in it. You can do this by selecting known safe brands[47] or by avoiding any products that list plastics in the ingredients list. Polyethylene and polypropylene are the most commonly used microbead plastics, but there are more, so avoid products containing nylon or any ingredient starting with poly... Homemade tooth powders can be made from food grade ingredients – which are undeniably healthier than any products that carry a warning not to swallow the contents.

I've had trouble with my teeth over the years. I'm ashamed to say that this may possibly be in part due to an overzealous desire to re-use dental floss – it sounds a bit excessive until you consider that we reuse our toothbrushes all the time! (My tooth health is also a legacy of copious amounts of sugar in my wilder younger days). Thus interdental brushes and antibacterial gel are currently part of my nightly scrub, along with a drop of Ratanhia tincture on my toothbrush and occasional oil pulling. Even with these purchases I can minimise the use of products to delay buying replacements. Hand in hand with minimising the use of products to help keep my teeth clean, I've pursued the less-stuff approach by shifting my diet to keep my teeth strong and healthy from the inside, guided in the main by reading *Cure Tooth Decay Naturally* by dentist Ramiel Nagel.[48] Happily, since starting into this process, I find that my gums and teeth are healthier than ever before.

Sanitary products and toilet paper

Sewage treatment plants are often ill equipped to filter out tampons and their wrappers and all the other bits of plastic thrown into toilets; the condoms, foil wrap, wipes, cigarette butts, toothbrushes, bottle caps and more. So many end up in the sea. Here they break up into microplastics and are seen as food for birds, fish and sea mammals. Natalie Fee of City to Sea works to stop plastic pollution at source, rather than attempt the impossible task of shutting the stable door after the toilet is flushed, so to speak.

She shares a pretty shocking TED talk[49] about the impact of plastic pollution in our oceans, and the research findings that 7% of all the plastic found on our beaches comes from our toilets!

Instead of the usual products, look for sanitary products made from recycled, unbleached paper. It's also easy to buy washable organic cotton instead of single-use paper products. Mooncups are soft silicone menstrual cups which can be washed and reused over and over again. Cutting out plastic pollution from our toilets will make a huge difference to marine life, and indeed to all of us who depend on the natural world for our very survival.

Regarding toilet paper, global use varies from 3.4kg/p/yr in Brazil to 12.7kg/p/yr in the US.[50] For the US this equates to 15m trees, almost 1.8bn tons of water in the manufacturing process, over 250,000 tons of chlorine for the bleaching process and over 17 terrawatts of energy in the manufacture alone.[51] If we were to opt for recycled unbleached paper at least that would remove the damage done to woodland habitats as well as reducing the energy and water footprint of our toilet paper and completely cutting out the need for toxic inputs of chlorine. If you want to go a step further consider a bidet or flannel wipe instead of single-use and flush.

Cosmetics and beauty products

People are beautiful. We don't need to dye our hair and paint our faces to be so. We need nothing more than fresh air, exercise and a smile to bring vitality into our features. Advertising agencies and film studios paint an unrealistic image of beauty that is only achievable with cosmetics and photo editing. Hair goes grey and thins. So what? Hiding this is part of our collective denial of the fact that we age and will eventually die. That will happen sooner and more painfully if we don't sort out the mess we're in on the planet.

However if you want to use makeup as a form of creative expression there are lots of low-waste ideas to work with. Generally, products from your local health food shop will be more natural, have a lower embodied waste and are less likely to be tested on animals. Not only are they Earth-friendlier, they're people-friendlier too. Why put something on your skin that you wouldn't eat? Either way it will get into the body.

Take only packaging that is easily recyclable or reusable where possible. A growing range of eco-friendly options is coming on stream, but buy with care to avoid greenwashing. Homemade cosmetics are another option. With a handful of natural ingredients, you can make foundation, toner, deodorant, eyeliner, mascara and other beauty products for yourself.[52]

Clothes washing

For clothes washing there are many environmentally-friendly alternatives available. As with shampoos, soaps, cleaning products and cosmetics, good health food shops carry washing powder or liquid that is less damaging to the natural world than conventional alternatives. Some laundry magnets/balls/discs seem to work well as an alternative to washing powder, or at least as a way to minimise powder use.

Some other considerations: a washing line is a low waste and low energy alternative to a drier. There are recipes online for detergent made from horse chestnuts. I've yet to try them, so would appreciate any feedback from those who have! You can buy stain remover as a bar rather than in a plastic bottle, or try rubbing some very soapy water into grease stains ahead of washing, so that you eke out the last of that stain remover bottle already in the cupboard.

Clothes washing is also one of the largest sources of microplastics in the world. Every time artificial fibre clothing such as nylon, polyester and the like are run through a wash they shed literally hundreds of thousands of individual microplastic bits. These can end up being eaten by zooplankton, the tiniest animals in the sea, near the very bottom of the food chain. When they do, they will inevitably be eaten by other animals and so on up the chain. With humans at the top of the food chain, that's a lot of plastic floating its way into our diet.

Plastics are already causing the death of ocean wildlife, so to cut out microplastics from your diet and everybody else's, take direct action: avoid buying synthetic fibre clothing and wash the synthetics you already own less often. Advocate – tell clothing suppliers to stick to natural fibres and tell governments to require microfibre screening on all new washing machines and on all sewage discharges.

Water

Excessive water use is one of those hidden wastes. In moist climates like the one I'm typing in right now, water is present in abundance. Here in Ireland we have a reliable source of water falling from the sky one day in every two on average. But this past summer (2018) saw some of the most severe droughts ever. Now, an Irish drought is a laughable thing if you live anywhere else in the world. The grass is still green, but the growth rate has dried up, so for our farmers and their livestock it's serious. With climate breakdown, this may well be our new normal.

However there are other reasons to save water. The energy needed to pump and filter water, and then to treat it after use, is vast. Paul Hawken's *Drawdown*[53] lists fixing leaks in mains water supplies no.71 in the ranked solutions to reverse climate breakdown. So, if fixing leakage can be so effective, surely we can help out by cutting down on usage within the home as well.

There are lots of the usual tips: turn off the tap when doing your teeth. Use a bucket and brush rather than a hose when cleaning your car or your windows. Collect and use rainwater for outdoor jobs and gardening. If you want to go a little further you can reroute rainwater to the toilet cistern, use a bucket in the bathroom for surplus clean water from hot water bottles etc. to flush the toilet, or better still, use a compost toilet and cut out the flushing entirely.

Don't forget that much of our water footprint is actually spent on our behalf on farms in arid regions for growing food such as almonds and avocados; and in factories processing clothing and other products for shipping it to our rain-drenched oasis perched in the Atlantic... So as always, be mindful of what you buy, live simply and keep an eye on petition websites to ensure that the businesses that serve us don't do so at the expense of other people and the natural world.

Nutrients and biomass

The waste that most commonly occurs in our bathrooms is not something that shows up in the bin at all. It's what we flush down the toilet. Every day we flush away gallons of clean water along with nutrients that could easily be used as a fertiliser; and biomass that could enrich soils and sequester carbon from the atmosphere (by converting CO_2 to plants, to food, to faeces, to stable compost in the soil).

Compost toilets are an excellent way to recoup these resources and avoid the waste and pollution that flush toilets generate. Hybrid technologies such as faecal separators and urine diversion toilets can offer the advantages of resource recovery along with the familiarity of flush toilets. These methods are the subject of my earlier books *Septic Tank Options and Alternatives* and *Permaculture Guide to Reed Beds*.

At the very least, consider using recycled paper toilet tissue. This is readily available and less overtly wasteful than felling forests and chlorine bleaching paper just to wipe your bum! If you can buy it in bulk you can reduce the packaging that comes with it.

The Bedroom

Here we look at clothes and bed linen, as well as footwear, storage and other stuff to do with the bedroom.

Clothes and bed linen

Fast fashion is every bit as challenging for the planet as fast food is for your body. The pollution load, the water footprint and energy use associated with clothing is huge.[54,55] It's estimated that £30bn worth of unworn clothes sit in UK closets.[56]

To avoid being part of the madness, buy only what you need, rather than routine impulse buying. If you stick to classic styles, the changes in trends and fashions will be less likely to 'date' your clothes. Select natural fibres for preference so that you don't send microplastics down to the local river with every single wash. Organic clothing has the added advantage of minimising embodied waste and environmental damage from spraying operations. Cotton is usually a heavily spray intensive crop, so investigate organic cotton, unbleached linen, wool, bamboo, soya or hemp clothing as ecological alternatives. If you buy good quality, lasting items they will keep serving you for longer.

Buying locally made clothing and bed linen can cut out some of the energy used in transportation, depending on where the raw materials were produced. Bring unwanted clothes to second hand shops to minimise waste and to keep good clothing in use. Older clothes are still valuable for textile recycling or rag recycling. Buy what you can at second-hand shops to minimise packaging and embodied waste in new clothes production and to keep your clothing budget as close to carbon neutral as possible.

Footwear

Shoes, boots, wellies, sandals, clogs, overshoes, waders, flippers, roller blades... Whatever you have on your feet, just use it carefully, care for it, repair it as needed, and use it out! Generally the more natural the process (real leather, rubber or cork) the less the embodied waste involved. Send good but unwanted footwear to second hand shops or free to a good home forums. Ideally find what you need in a second hand shop to cut out the resource use inherent in a new pair.

Barefoot summers are a good zero-waste option, with the added benefit of electrically earthing you at the same time which apparently is good for optimum health[57] (so you can save on medication waste too).

Eye-wear

Whether you wear glasses, sunglasses or contact lenses, there is an eco-footprint to consider. Contact lenses produce the most day-to-day waste, but the acetate manufacturing process for glasses isn't all that great either.[58] Care for what you already wear, replace the lenses if you need new lenses rather than changing the frames as well. I think I've had mine for over 20 years, which is ten times the average,[59] and going strong.

Avoid the 2 for 1 offers for the obvious waste reasons and also because when you look at them closely they usually get you to spend more money anyway. If you do change your specs, let the optician route your old pair to a new home through a suitable charity. If you want to tackle this one from another angle and get rid of glasses entirely look up Martin Brofman's *Improve Your Vision*.[60]

Storage

Storage can be a problem area in many houses. It may not seem like a waste issue per se, but just think of the amount of waste generated by building an extension. Most extensions are built because 'there isn't enough space for everything!' One immediate way to minimise on storage requirements is to minimise the amount of clothing (and other things) that you have. Check what you actually use, and keep that, but send the rest off to a charity/thrift shop so that somebody else can enjoy it other than the moths.

Contraception

Some methods of contraception are closer to zero waste than others. Fertility awareness or Natural Family Planning (NFP) methods have the plus sides of being free of waste, wrapping, devices, drugs or hormones. They're not an on demand method though, so not necessarily every couple's cup of tea.

Hormone-based contraceptives such as the pill, the patch and hormone-based IUDs (Intrauterine devices) have the potential to contaminate the natural world via our urine.[61] Endocrine disrupting chemicals can impact on freshwater fish as well as grazing animals fed in fields with sewage sludge applications, and on people via our drinking water.[62] As little as 5 nanograms per litre of the steroid used in the pill can cause negative effects in fish[63] and they're not necessarily the healthiest pill to swallow either.[64] These are reasons enough to explore the many alternatives.

A quick internet search will show up options such as Fairtrade or organic condoms (to avoid unwanted chemicals),[65] copper IUDs, surgery and more. The decision will most likely hang on more than just your zero waste endeavours. Clearly any product used regularly will have regular packaging waste. However, bear in mind that the more children we have, particularly in western consumerist cultures, the greater our overall ecological footprint – by a very large margin. So whichever method you select; however many children you choose to have; however utterly loveable a happy accident can be, from an ecological standpoint any packaging waste generated by contraception is probably worth it.

Keeping warm in bed

The previous section might give you ideas for keeping warm in bed, but if not, then some other options are electric blankets, hot water bottles or to heat your house to the point where the ambient temperature negates the need for any of the above. The last option won't generate waste in the house, but the ecological footprint of home heating is considerable, so that's not necessarily the best route to take. Electric blankets should last longer than regularly used hot water bottles, but are a bulkier waste that will need to be disposed of as waste electronic equipment due to the electrical components present. Recently we found an old ceramic hot water bottle in an antique shop and it's been serving us well so far. The old rubber seal had long since perished so I cut a few washers from a bicycle tube to replace it.

If you are keen on water saving, a bucket in the bath can serve as a useful way to recycle hot water bottle water for toilet flushing.

The Living Room

Games, television, reading and music can all play their part in contributing to waste, or contributing to the solution instead.

TV

Television is a disaster for waste minimisation. There is no other medium through which the message of 'buy more stuff, buy more stuff, buy more stuff' comes with more force than via TV. To be attractive, popular, sexy or intelligent buy this or that latest product. Whatever your imagined character flaw happens to be (and comparisons with the rich and famous, made over and photoshopped are a sure way to find it) the message is that if you buy something then you will be OK. Until the next ad, soap, or film comes on.

The average TV watching time in Britain and Ireland is about 24 hours per week, rising to over 30 hours per week in the US. That is a lot of ads telling us and our children what to think, what to do, what to say, what to eat and most importantly from a waste point of view, what to Buy. Lots! is the constant mantra. We haven't even got to mentioning DVDs, computer games, streaming services and all that. Basically the mainstream productions urge greater consumption, even if it is travel shows showing beautiful people swimming over coral reefs. The ad there is flights, hotels and foreign holidays.

Suggestion No.1: get rid of your TV. Dump it, or give it away to a friend (or enemy if you have one) or a charity/thrift shop or anywhere, but get it out of your house. If that is a bit extreme for you, then drastically reduce the amount of time you watch it, and be very selective about what you and your children watch. You don't need a TV to live life to the full, and getting rid of it greatly helps in the process of living simply.

Games

Maximise versatility: a deck of cards offers far more opportunities than a big box board game, for example. If you go for natural materials you'll find it easier to recycle than if the game has lots of plastics and electronics. Avoid rushing out to buy the 'very latest thing' if you think that it will not last the pace. There are as many trends in toys as in clothes. Beware of them. Trends come and go and can leave the residue clogging up your house afterwards.

Avoid toys with batteries to minimise this hazardous waste source. Old toys can be given to charity/thrift shops if they are in good order, or sold on LETS (Local Exchange Trading Systems) or car boot sales. Make full use of available materials, particularly for young children. Cardboard tubes, wooden spoons, saucepan lids and other everyday objects make a fascinating world of wonder for toddlers and young children. Washing up at the kitchen sink is often far preferred to owning a mini plastic kitchen.

Reading material

Newspapers are readily recyclable. They can also be soaked in water and used as a thick mulch around young trees before being covered with grass clippings. They can be cut into strips and soaked in a white-flour paste for papier-mâché projects. They can be used to light the fire in the mornings. However it still takes energy and resources to produce and transport the paper in the first place. How many do you buy each month? Can you halve that amount?

Magazines are often glossy, and are thus not suitable for garden mulch or papier-mâché and should not be burned. They do recycle though, but recycling paper takes energy too, so minimise your purchasing if possible. Most environmental magazines now come in recycled, unbleached paper: look for these in preference to virgin, chlorine-bleached paper. Publications with an ecological, social or global consciousness are also more likely to confirm your ideals of waste reduction rather than selling you the opposite.

Used but good magazines may be welcome in the waiting room of your local health care practitioner. To get twice the reading for no extra resource use you can swap magazines with a like-minded neighbour. For books, the library is an invaluable resource. Not only is all the existing stock of books at your disposal, but sometimes you can ask them to order in new books too. Some second-hand bookshops will do a part-exchange on returned titles.

Music

Music can produce waste in the form of LPs, tapes, CDs and old stereo equipment. If you thought that streaming was the answer to everything, think again. Streaming online content, be that music, films or whatever, takes a lot of energy to power the vast server farms now used to handle the quantity of data being generated. According to John Harris in the Guardian, all data streaming is projected to consume more energy than the entire aviation industry.[66]

Make your own music if you have instruments, or sing. Children in particular love this interaction. Listen to the radio or try your library rather than buying more CDs.

Appreciate the music you already have, and buy carefully if you do want more music around. Take good care of your musical equipment and instruments, so that they will last a lifetime. Find a local singing club or local orchestra. Form a band. Jam with friends. The opportunities are endless.

Lighting and Heating

Lighting and heating are significant sinks for energy and sources of waste. The introduction of domestic low energy lighting is 33rd on Paul Hawken's

Drawdown list of actions to reverse climate breakdown and 44th for application in commercial buildings and cities.[67]

Hawken lists district heating, insulation and solar hot water all amongst the top 80 measures to reverse climate breakdown. Clean cook stoves are listed as the 21st most effective measure (in the context of replacing inefficient open fire cooking facilities rather than a measure for most of our modern kitchens.

Light fittings and bulbs

Usually low energy is synonymous with low waste. Unfortunately this isn't the case with light bulbs. CFLs may last longer than tungsten bulbs and have a lower energy usage, but they are fluorescent tubes with an unhealthy light spectrum[68] and contain mercury. On both medical and ecological grounds they are worth giving a wide berth. If you have them in the house consider replacing them steadily as they blow or give them away to somebody who will otherwise buy new ones.

LED bulbs use less energy again and last longer. However they also contain heavy metals[69] and thus contribute to household hazard waste. Also, the light spectrum from LED is in the blue range, so they are best used during sunlight hours and avoided in the hours before sleep.[70]

Halogen bulbs (a tungsten filament set into a bulb filled with inert gas and a halogen gas such as Bromine or Iodine) use more electricity, but in our own household we've migrated back to these (while we still can)[71] to get a more complete light spectrum in our home. This is hand in hand with buying our power from a supplier of renewable electricity to limit the impact of our energy consumption. Both CFLs and LEDs contribute to hazardous waste[72] at end-of-life, whereas traditional tungsten bulbs are safer for manufacture and disposal/recycling as well.

Certainly filament bulbs are less energy efficient per lumen emitted, but where the heat generated becomes part of the space heating budget that efficiency may be exaggerated. Another common argument against filament bulbs is the short lifespan, however the longest lived filament bulb has been burning pretty much uninterrupted since 1901.[73] As we begin to grow up as a society and move beyond planned obsolescence,[74] I'm sure there will be plenty scope for innovation and the development of a bulb that is efficient, durable and yet non-toxic.

Bulbs of any sort are essentially single use items. Technically CFLs have the mercury gasses removed and LEDs should go to the electronic waste section of the recycling centre, but given how much WEEE actually gets recycled perhaps we'd be a lot better off designing our homes and commercial buildings to allow more natural light. In fact, on grounds of health, energy and waste, that would be the very best option.

Heating systems

Heating systems come in all shapes and sizes; fuelled by oil, gas, solid fuel (wood, coal, peat), electric, solar, or a combination of these. Oil, gas and electricity are all low producers of on-site waste. However they are all fossil energy derived, unless you are signed up to a supplier of renewable electricity. Thus they use up scarce, non-renewable resources, contribute to climate breakdown and should be avoided where possible. Coal has the further waste-related drawback of having coal bags and coal ash to dispose of afterwards. Avoid using coal ash in the garden due to the presence of heavy metals. Peat, another fossil fuel, comes from our beautifully diverse peatland habitat, a valuable carbon sink, and should be avoided too.

If you're not already signed up with a green electricity company, then stop reading, look online for a supplier of 100% renewable electricity and make the switch immediately! That said, the simple truth is that if we remove dirty fuels from our electricity supply networks and continue without aiming for energy reduction too, there won't be enough power there to meet the demand. So we need to both switch to renewables, hand in hand with dramatically reducing our overall use of electricity.

Timber can be ordered by the trailer load if you have somewhere to store it and can be replanted as soon as it is cut, ready to harvest again in less than a generation. Coppice management is even quicker, allowing harvesting of wood for firewood every 5-10 years or more depending on the variety of trees being coppiced. Wood ash can be used around trees in small amounts as a mulch and fertiliser. It is high in potassium, which is good for fruit trees and bushes.

Do not burn plastic in the fire since many plastics are toxic when burned. That includes the envelopes with windows, 'cardboard' cartons, plastic tape on boxes, plastic in floor sweepings, firelighter bags, peat briquette straps and anything else that may be tempted to cross the hearth. Avoid the plastic to protect the natural world, your health and that of your neighbours.

Solar heat is readily available and should be harnessed where possible. South facing glass, solar panels and careful house designs are all good ways to maximise catching the sun's rays. Houses can now be built to passive standard, requiring almost no external heat sources – after the sun, cooking and body heat are considered.

Pellet stoves, wood chip boilers, log gasifiers, geothermal heating and solar panels are all increasing in diversity and efficiency each year. For an in depth look at home heating investigate these options thoroughly. PV (photo voltaic) cells are sometimes connected to an immersion to convert the excess electricity into hot water directly rather than feeding it back to the grid or to batteries.

Fireplaces

Stoves are much more efficient than open fireplaces and can easily be retrofitted. If you are at the design stage for your house investigate masonry (ceramic) stoves or rocket stoves for maximising fuel burn efficiency and minimising fuel use. *The Green Building Bible*[75] has a chapter on wood burning. Patrick Whitefield's *Earth Care Manual*[76] has a section on heating with wood, including a description of masonry stoves, as has John Seymour's *Complete Book of Self-Sufficiency*.[77]

From a waste perspective, the more efficient your heating system, the less ash produced in the year, and the less effort and resources devoted to providing the wood fuel. If you are using wood, the ash can be used to enrich your fruit trees and shrubs. Consider, however, that wood is not deemed to be a smokeless fuel in some areas, so may not be an option where you live. This is sometimes possible to overcome if you use an approved stove that gives a cleaner burn.

Furnishing and Appliances

As with so much else in our society, we make too much furniture and too many appliances, of increasingly poor quality, with an ever shorter life expectancy and limited opportunities for genuine reuse and recycling. However as always, there are many ways that we can minimise our impact whilst still meeting our needs.

Fitted furniture

Fitted furniture is generally less versatile than freestanding. However if you are putting it in, design it for maximum re-use elsewhere so that it can be moved to another room, a new house, sold, or given away rather than dumped. Otherwise try to reuse the materials when they are at the end of their life instead of dumping them. We've seen plain 9in x 1in (22.8cm x 2.5cm) timber shelves take many incarnations, changing from room to room as the family has grown and our needs have changed. If you use natural untreated wood rather than wood composites, painted or preserved wood, then the timber can be safely used as firewood after it is finished with, if no other use can be found for it.

Freestanding furniture

For other furniture minimise chemical inputs and maximise the use it gets, either in your own home or in somebody else's. Local free ad newspapers, free to a good home forums and Freecycle are wonderful vehicles for sending on old furniture to new homes or finding new furniture for your own home. Use natural timber rather than composite-board products. Source locally made native hardwood timber items instead of tropical hardwoods. These can often be found in small specialist furniture shops. Remember that second-hand items are essentially carbon and resource neutral.

Floor coverings

For floor coverings remember that different finishes have different energy inputs, waste outputs and lifetimes. Look for products that are natural, durable and easily recyclable or at least safe to downcycle into firewood or aggregate.

For minimum waste generation in the long term consider stone, quarry tile or timber that will have a good resale value when it is removed if the house is being renovated or demolished. If disposing of existing floor coverings: carpet tiles are very transportable, carpets can sometimes be reused or passed on to a suitable home, or (if made of natural materials) used to cover vegetable beds as a weed suppressor.

Soft furnishings

As per clothing, source soft furnishings made from natural materials where possible, to avoid microplastics coming loose in your washing machine. Remember that if you can find the materials or items you need in a second hand shop, antique shop or free to a good home forum then your resource use will essentially be nil. DIY and repair both offer the potential to keep materials and furnishings in active use, or can be downcycled into something else (usually of lower value than the original furnishing – paradoxically you'll find ideas for these under 'upcycled furniture' on a web search).

White goods

If you are buying a new fridge or washing machine, find the most energy efficient and water efficient model you can. This will minimise your energy consumption and the related waste involved in electricity production. Before buying, however, ask yourself if it is necessary in the first place. An old model can be more efficient than a new one if you count in manufacture and transport, energy and waste. Old fridges and freezers need to be taken to the retailer or recycling centre for refrigerant gas recovery. Do not forget second hand shops or free to a good home forums if you are getting rid an old but functioning appliance. Remember that great care must be taken in transport to avoid damaging the item, leading to a leak of refrigerant gasses.

When the Drawdown Project in the US researched the measures that could be taken to reverse climate breakdown they looked at reduction of greenhouse gasses and also carbon sequestration measures for removing carbon from the air again. The number one item on their list was sourcing more climate-friendly refrigerants.[17] The HFCs refrigerant gasses used in fridges, freezers and air conditioners have a monumental greenhouse gas potential of thousands to tens of thousands of times that of CO_2.[78] If you are looking for a new appliance that has a heat exchanger, then ask for climate friendly alternatives to these gasses. More and more manufacturers are moving over and every time you ask it helps create demand for more climate friendly products.

Children

Apparently, the biggest single thing you can do to reduce your impact on the climate is to have fewer children. According to a 2017 study by Seth Wynes and Kimberly Nicholas in Environmental Research Letters[79] this can reduce your carbon footprint by as much as 58.6T of CO_2-eq/yr, versus the next nearest item on the list at 2.4T/yr for living car-free. In this context, educating girls and family planning have been listed in Paul Hawken's Drawdown as the 6th and 7th most effective ways to address climate breakdown.[80] Of course, where you live and what lifestyle you lead will have a huge bearing on these figures. The same basic principles apply to your waste footprint.

The original third ethic of permaculture, along with Earth Care and People Care, was 'setting limits to population and consumption'. It seems that this was quite a controversial phrase, which has led to a host of variations over the years,[81] most recently Fair Shares or 'future care' depending on who you talk to.

My early memory of population control arguments seemed to be most vociferous coming from countries with the highest consumption levels – without any mention of reducing that consumption. It seemed to me then, as it still does, that pointing an accusing finger at countries with modest economic means and high birth rates simply wasn't the answer. We need to re-embrace that third ethic again and acknowledge that the carrying capacity of the planet is indeed limited, and that our impact equals population size multiplied by consumption per person.

Happily, we don't need to rush out there with rules, regulations and restrictions to get the population under control. Swedish statistician Hans Rosling's TED talk[82] offers a very encouraging perspective on the reduction in birth rates that results quite naturally from education and economic stability. We do, however, need to regulate consumption. A relatively easy way to do that is to introduce Cap and Share on fossil fuel extraction (see the Legislation and Government Policy section).

If you're fortunate enough to have children in your life, here are some of the common ways that you can still aim towards zero waste.

Babies: nappies, creams, shampoos, clothes
Nappies are the waste that is mentioned over and over again by parents wishing to reduce their bin weight. The easy solution: use cloth nappies. These are available now in fitted shapes with Velcro fasteners, comfy waterproof covers and paper inserts. The inserts were the thing that won me over – just lift up the contents and flush down the loo! Certainly easier than a generation ago. Better yet, put them in your compost loo and build rich soil.

Single-use anything is a bit of a disaster, but if you really feel you can't work with cloth nappies or need disposable ones for travelling etc., then recycled paper disposable nappies are available in good health food shops. These are much more eco-friendly than the conventional, chlorine bleached, virgin paper equivalent.

Instead of medicated creams, you can buy natural calendula cream in glass jars or metal tubes. To get out the last drop before dumping tube packaging, roll the tube on a bread board with the handle of a wooden spoon. Shampoos in health food shops are usually gentler on your skin than supermarket or pharmacy equivalents, so perhaps your refillable shampoo bottle will do for washing babies and infants as well as you.

Charity/thrift shops, dedicated second-hand sales for baby equipment and clothing, and friends are great for sourcing babies' clothes because they grow out of them so quickly. Then send the clothes back to these sources again afterwards to keep them in circulation, staying out of landfill and avoiding the need for making more in the world.

Older children: entertainment, clothes

Entertain without battery operated toys where possible. Maximise your time input with your children instead of stuff input. Use charity/thrift shops, second hand shops and online forums to both get and get rid of toys and clothes. We had a general rule that said: "OK, if you can find a bag of things for the charity shop, you can buy something there to take home", thus getting the children's decluttering muscles going and instil the value of using pre-loved items.

Teens: entertainment, clothes, make-up?

As children get older, the same ideas apply. Again, maximise time input over stuff input. Let them see the waste issue in terms of what they buy and how they then have to pay for its disposal (financially and ecologically) if it is not routed elsewhere or avoided on day one. Most teens are more clued into the multiple environmental crises than their parents, so this may not be a hard sell.

Fast fashion rules may apply here more than to your own wardrobe since teens are in the first flush of having some money, but not always enough to select higher quality, more durable and classic clothing. Treat your teens to some (diplomatic) guidance and possibly a supplementary budget to steer them towards clothing that will last the pace rather than encouraging them to participate in the cheap clothing race from sweatshop to retailer to incinerator. Encourage them to be aware of the social and environmental problems associated with fast fashion.

As for make-up; as children grow into adults, they will experiment with styles and with their way of presenting themselves to the world. Don't let waste minimisation become an obstacle to good relationships – that would be a waste indeed. If they are so inclined, let them experiment with home made cosmetics and make-up. If not, don't worry too much. This is unlikely to be the largest source of waste in your home.

School
Perhaps the trickiest source of waste is the stream of new school books every year. We have remarkably little input into the selection of books required for our children at the start of each school year. However we can write to our schools and Department of Education encouraging them to adopt school book rental schemes and to reduce the high turnover of books selected for the curriculum.

Copy books, pens, pencils etc. can all be sourced from a supplier of recycled stationary. Look for unbleached or oxygen bleached, 100% post consumer recycled paper products. The standard process of using virgin wood and chlorine bleaching causes untold trouble for the forests and aquatic habitats where these are carried out. Unpainted wooden pencils and cardboard casing for pen inserts all reduce the ecological and waste footprint.

Lunchboxes can be used instead of cling film (PVC plastic which can leach chemicals into our food) or aluminium foil (which is very energy intensive to manufacture, is not readily recyclable via commercial collection facilities, and is neurotoxic[83] so shouldn't be in contact with your food). Alternatively use greaseproof or parchment paper which can be washed and dried for several uses, or get multiple uses from a small plastic bag, or use a beeswax wrap.

School can be a great source of friends, fun and stretching of the mind. But if the messages communicated reinforce that of a consumer society, then they can have their limitations too. In what has become the most popular TED talk of all time, Ken Robinson asks if schools kill creativity.[84] Now more than ever before, we need to be creative. The multiple issues of ocean acidification, wildlife loss, species extinction, climate breakdown, water pollution, soil erosion and a myriad of chemical threats to our health are pressing in on all sides. As a matter of urgency, our education systems need to shift away from teaching us how to fit into a company job to prop up the status quo and towards the skills that will help us survive the coming decades and century ahead. We need creativity, adaptability, social skills, cooperative skills, critical thinking (to see past the false promises from advertising and politics), self-reliance, education around what foods are healthy and which ones are not (hint – if you shop in supermarkets and emerge with a trolley full of plastic, then the contents are less likely to be good for you).

Parties and Festivals

Looked at from afar, the Earth is blue. Closer up, it's a beautiful dappled collection of greens, blues and brown hues. However if you come just a little closer in, it looks like one big party, with a little too much noise and a little over-inebriated. Since we've discovered how to convert oil into food (which we're doing via the industrialisation of agriculture) we've done what any colony of bacteria on a new petri dish will do. We've partied. We've reproduced abundantly and lived as if there was no tomorrow. However, once the nutrient supplies in a petri dish run out, the population of bacteria crashes. That's not something that we want to have forced upon us here on our beautiful blue planet. Perhaps it's time we learned how to party without taking quite such a toll of the carrying capacity of the Earth.

Birthdays

Where to start?! Keep parties activity-centred, rather than stuff centred where you can. If you minimise the unhealthy stuff, the crisps, sweets, chocolates, biscuits etc., you will generally cut down on much of the plastic wrapping too. Health, wallet and environment go together here: home baked savoury party foods may be enjoyed every bit as much as sugar-stuffed plastic-packaged sweets and snacks.

Gifts are also an issue here. Exercise your imagination to find gifts that will be appreciated and don't cost the Earth. Any number of environmental or conservation charities and eco-shops have sprung up over the last decade or so with gift ideas that have minimum ecological impact. If there are none in your area, check the internet for online eco-friendly products (and ordering as close to home as possible to reduce product miles). If you're stuck for ideas, gift vouchers or book tokens allow the recipient to select their own gift, which at least reduces the potential for yet more unwanted things in the world. Consider experiences as gifts.

Overall, it may work to gently tone down the gifts or to consider asking others to cut down on their gifts to you. How about a homemade card, breakfast in bed or a birthday hug?

Weddings, funerals and other ceremonies

Weddings, funerals and other religious or traditional ceremonies are all times when huge pressure can be brought to bear on people to have the latest this, that, or the other, or to follow tradition rather than personal preference or common sense. Use these occasions to put all the other lifestyle changes into practise.

At a minimum, at gatherings of any sort avoid using disposable cups, plates and cutlery. Better a slight mismatch of ceramic cups and plates than a bin full of single-use paper and plastic resources later on. Friends often weigh in on such occasions with offers of salads and dishes of different sorts; enjoy the communal element of these offers rather than dashing out to buy prepared packaged meals.

Weddings can be mad affairs that cost an arm and a leg at a time when money is often at its tightest. Although not always, it tends to be that the higher the cost, the greater the waste associated with it. Consider zero waste at every step of the way. Perhaps a beautifully designed hand-painted design can form the basis of an email invitation to your friends instead of a posted card. Instead of hothouse imported flowers, try locally grown, in-season flowers. Beautiful wedding dresses can be found second hand and returned after the big day so that the overall waste and energy budget gets smaller per person every time it sees a new wedding.

We had our reception in our garden and used stacks of pallets for tables and borrowed chairs, cutlery and crockery. We borrowed a small marquee from friends in case of rain. I was worried that our old septic tank wouldn't last the pace, so I built two compost toilets in the garden (making sure to follow good practice so that we didn't cause groundwater pollution). One privy building was of pallet wood and one of straw bales. By the time we returned from honeymoon, the makeshift straw bale building had fallen in and a year luter all that was left after I removed the tin roof, loo seat and some supporting blocks was composted straw and a rich pocket of soil in the ground.

Funerals can also be opportunities to step away from the norm and towards the more ecological option. When my grandmother passed away recently at the grand age of 92, her children looked for eco-friendly coffins. A woven willow option was considered but was imported, so at the end of the day a locally made pine coffin was chosen; plain, unvarnished and beautiful. You can also get cardboard coffins or burial shrouds that are much lower impact than rainforest teak with goldy handles.

If the thought of a long-term rental plot in your local graveyard or the energy inputs of cremation both sound unappealing, consider a natural burial instead.[85] Places such as the Sustainability Centre in Hampshire offer a peaceful final resting place marked by a new tree rather than a headstone. Imagine for a second if all of us opted for that – that's 7.5bn more trees with just that one change in how we approach both life and death. The option I hope is available when my time comes is composting.[86] Plant a tree over me by all means, but I don't want to be six feet down where the roots and other myriad of soil life forms can't easily get at me and return me to the earth properly.

Festivals and family events bring out everybody's expectations. One of the challenges is the peer pressure to do the normal, high cost, high waste thing at every occasion. When it comes to waste, climate and resource use, however we continue on our current trajectory at our peril. Besides, the most enjoyable and memorable events are often those that threw convention out the window and tried something new.

Christmas, Easter, Hallowe'en

Are these festivals a childhood dream or a waste minimisation nightmare? It is possible to celebrate the year without enormous waste generation, but it takes a lot of thought, commitment and imagination and oodles of tact. It can be a challenge to control what actually comes into the house at these times without causing offence to all your in-laws and out-laws, the Easter Bunny, neighbours and friends. However you can ensure that the presents you give contain loving intent rather than unnecessary stuff bulked up with pretty packaging.

As an example, charity donation cards provide a way for giving goodwill rather than things, in a way that actively helps those in need. If you have a favourite supplier of pretty recycled stationary, eco-friendly kitchen cleaning products or even silk dental floss and health-foody body-care products, then Christmas is your chance to increase the eco-friendly average in your gift giving; and advertise your favourite suppliers at the same time.

Easter can be a time of making chocolate eggs rather than buying the packaging as well. If a bought egg is still desired by your household, shop for less packaging and recycle all the plastic and card that comes with them. An Easter egg hunt in the garden can be a fun way to start the day and can distract from the fact that the eggs themselves are the low-packaging option this year. As children, we started our Easter Sundays with a boiled egg that we painted before cracking it open to eat it (before cracking open the chocolate eggs, it must be said).

Hallowe'en is traditionally associated with nuts, dried fruits and apples, all of which have lower waste implications than most chocolates and sweets. Local trick-or-treaters may not be impressed with some fruit and nuts, but your own family might just permit a slight shift from highly packaged sweets to a more diverse range of eats. Look for sweets that are easy to give out at the door and yet are low in plastic wrappers. These won't necessarily be the cheapest sweets on the shelf, but it's worth looking locally to see what's available. Try giving out biscuits instead – one wrapper is better than many small ones. If you are good at home baking and have the time and space to run something up, that can be a good way to reduce Hallowe'en waste.

Music festivals

I confess I'm not a festival-goer... There, I've admitted it. Greening of festivals is now a thing though, so it's easier and easier to avoid the waste involved. Festival organisers have lots of potential for cutting down on waste and requiring traders at their events to do likewise. Once you're through the gate as a participant, your choices for food, drink and sanitation plummet, but you can write to organisers and encourage them to keep their events plastic-free and to provide composting collection for plant-based plastic containers and cutlery.

Some pretty simple things can make a big difference. Keep your tent afterwards, for example, rather than simply leaving it in the field behind you. It's strange to have to say it, but given some of the post-festival footage out there I thought I'd risk causing offence and put it in here as a reminder. Bring a picnic if you can, so that you minimise the packaging inherent in festival food. If eating in, look for stalls that sell food without packaging, or with basic wrapping such as a paper bag or napkin only. If composting facilities are available your paper can go in there (not in with recycling unless it is clean).

As regards sanitation at large festivals, l'Uritonnoir[87] is a French-made urinal which fits into a large round straw bale. Convert beer into fertiliser by simply passing it through the male portion of your festival-going population. Who says that composting can't be profitable? Ladies – there is a range of portable urinal devices on the market,[88] plastic, but arguably better than chemical loos. Perhaps festival organisers could take a lead from Dutch designer Marian Loth, and include the Lady P women's urinal[89] into their festival toilets and sell the liquid gold as an agricultural fertiliser.

Gardens, Sheds, Pets

There are as many opportunities to practice zero waste outside the house as there are inside. Here we explore garden maintenance, garden furniture and machinery, landscaping, pets and transport.

Trimmings/prunings/mowings/weedings

Don't dream of binning your garden trimmings, prunings and mowings and the nutrient value that is in them. Just make compost in a quiet corner of the garden. Bigger branches can be cut and dried for firewood. You can mow small trimmings so that they are mulched for the compost heap, but if you're not sure about the capability of your mower then you can shred them if you have a chipper or simply pile them up in a secluded corner and let time do its work. While you probably already have a container for kitchen compostables, the garden compost heap need not be such a secure process. Rats and mice won't be all that attracted to your hedge clippings. Build the heap in layers so that mowings don't make it all too sludgy, or trimmings make it too dry and airy (see Chapter 7 on composting).

Landscaping and plants

When designing hard landscaping, maximise the use of naturally available stone, timber etc. in preference to materials that will need to be disposed of when the landscaping changes. Keep it simple. Use materials already on-site where possible.

For soft landscaping, you can grow plants from cuttings and seed to reduce the packaging and cost of pot grown plants. Friends' gardens are often abundantly planted with raw materials for cuttings. Swap, for twice the minimisation. Again, keep it simple. Avoid peat moss to save the bogs and also coconut fibre, due to the product miles. You will have plenty homemade compost. If you really cannot get to making it yourself, find a source of good local compost or well rotted manure. Check that manure has plenty worms, so you know it isn't bringing vermicides (dewormer medication) into your garden.

As always, minimise the packaging you buy. Instead of endless seed packets with a mix of plastic, card and foil, consider saving your own seed. Many places now have seed and plant swaps where you can bring what you have in abundance and leave with an armful of everything else.

Alternatively buy seeds in plain polythene sachets (which can be reused with your own seed at the next seed swap) or better yet, in plain paper envelopes that can be reused, recycled, composted or burned. The aptly named Brown Envelope Seeds in West Cork offers just such a service.

There are lots of ways to reuse things in the garden. Instead of buying rubber tree ties, for example, old bicycle tubes can be cut into strips and hammered into your tree support stake instead.

Poles, pots and pieces of string

I'm writing this as the summer approaches. My annual search for bean poles and pea frames has begun. Pots are in short supply, having returned a hundred or so to my wetland plant supplier recently. We've just found that our baler twine, the staple for tying up tomatoes in the polytunnel, sheds microplastic bits all over the path as it decays. But as always, there are natural solutions to avoid creating waste.

Poles are available in abundance in our garden from the different osier willow cultivars I grow for new sewage treatment projects. Half a minute with a secateurs and I've a new 6-8ft pole ready to stake the runner beans. Willows are easy to grow from cuttings, so you need never be stuck.

Pots are best returned to the wholesale suppliers for reuse if possible. Otherwise reuse pots yourself or send them in to your local school or community garden project. If you find that you've run out, egg boxes and toilet roll tubes make an excellent standby, or make pots from newspaper by folding them carefully into the right shape. Keep cardboard or paper pots watered regularly because they dry out on all sides as well as the top.

If you have old pieces of nylon string already available then by all means use it up, but if buying new then jute, wool and hemp make natural, biodegradable alternatives. Keep an eye out for local or organic options. I was on a fire-making workshop with my elder daughter recently and the organiser had samples of strong pliable string made from briar and nettle fibres, so the options are present in abundance if you have the time, energy and inclination to seek them out.

Garden machinery

Garden machinery such as mowers, electric clippers, strimmers and the like, all take up a lot more space and use more energy than the manual alternative. To minimise the space they take up in your bin at the end of their lives, avoid buying them in the first place. Ask yourself; do you want the kind of garden that needs that much work? With a wild garden you reduce your carbon footprint, cut back on work and create space for pollinators, birds and other wildlife. Mow pathways and small lawn nooks and children will love it.

If you do want an array of garden tools, keep them well maintained so that they last as long as possible. Is it just my imagination or does older equipment tend to outlast newer models fairly consistently? The moral, keep your old mower and get it serviced and repaired rather than replaced.

You can also adopt a permaculture garden design and grow a forest garden where your lawn once was: no mowing, less watering, visually attractive and lots of fresh fruit and nuts in the harvest season. Patrick Whitefield's *Permaculture in a Nutshell*,[90] *The Earth Care Manual*[76] and *How to Make a Forest Garden*[90] can all help to guide the way.

Garden furniture

Make your own garden furniture from old timber, pallets, stone or old brick. Alternatively buy furniture made from naturally durable timber that can be used as firewood when it reaches the end of its life. Avoid preservatives and treated wood insofar as possible (see the 'problem ingredients' section in the composting chapter for pallet preservative codes).

Care for outdoor furniture appropriately to lengthen its lifespan. Over-wintering it in the garden shed is one way to help it live longer without chemicals. Seek out additives and preservatives with care. Boiled linseed oil sounds great, for example, but many brands can include an assortment of heavy metals or other toxins to speed the drying process. Raw linseed oil, by contrast, is natural, but is slow drying and may not be suitable for all applications (e.g. for outdoor benches if you wear your Sunday best into the garden). We drained out some large olive oil tins recently and now I have a jar of dusty olive oil. I'll try that on the polytunnel bench and let you know if it leaves oil marks on my gardening clothes. It has worked a treat on tool handles that were desiccating in the heat of the tunnel.

Shed storage

When it comes to storage, minimise your junk – that will minimise the storage space you need. Sheds are great for holding lots of stuff and are a lot less resource hungry than a house extension, but if you send unwanted stuff off for appropriate reuse or recycling then a smaller shed will do.

Moving things along also keeps them in use, minimising everybody's consumption of new, resource-intensive things. It can be easier to let things go if we reframe how we look at our possessions: could someone else be getting good use out of this thing I am storing instead of having it sitting in my shed/shelves/spare room? If the answer is yes, then perhaps we should pass it on and give it a new home. If not, then why are we hoarding it if it's no good to anybody? Either way, we can use this as a lever to clear clutter and free up shed space.

If you want to try your hand at a DIY building project, then a shed is certainly an easier one to start with than a house. Have a look at the section on DIY and construction for tips and reminders.

Pets

Feed pets on meat scraps as well as any other food they may need – make friends with a local restaurant and get good meat scraps from them rather than heavily packaged pet food. Keep hens rather than budgies as a way to use kitchen scraps and to minimise egg packaging.

It's no longer legal nor acceptable to let your dog foul the pavement or park. But if you wrap that little packet of biomass in plastic it will sit in a landfill for a thousand years or end up in an incinerator. Best to use biodegradable plant-based plastic bags for your dog do. These are available in good pet shops, or reuse the biodegradable bags from your local organic veg supplier. Ask in your usual shop if they don't carry them in stock already.

For cat lovers, there are lots of DIY cat litter options that are more eco-friendly than heavily packaged imported proprietary products. These are often readily available items such as newspaper, wheat, pine and cedar chips, wood shavings or dried orange peels. Ashes were the traditional material, but the pets then take the wood ash all around your house, which isn't ideal. Pet litter can be composted, but for safety it is best to keep this separate from the main compost area and bury the finished material away from your veg growing areas. It will still build soil and keep your waste load down.

DIY and Construction

Our homes are often the most expensive purchase we will ever make. They have a waste and resource footprint to match. Keep in mind that the smaller the home, the lower the resource inputs. If your budget begins to spiral out of control, chances are your waste footprint is keeping pace with it. Rein in, learn the value of voluntary simplicity, stay on budget and on course for a modest home and a tenable future where you don't find yourself working in a planet-wrecking job to cover an enormous mortgage.

The ultimate waste minimisation measure in DIY is 'don't build'. Shortly after we had our second child we decided not to extend! Compared to extending, it has reduced waste, resource consumption, heating bills, space-to-be-filled-with-stuff, financial outgoings and the stress of doing the job. Instead we decluttered, sending most of it off to charities, friends, recycling and bookshops. Suddenly we had a spare room again, instead of being a room short!

Of course, hand in hand with decluttering, another vital component in zero waste endeavours is to stop bringing stuff into the house – to simplify.

If your DIY skills leave a little to be desired, but you still have projects that you'd like to finish on an eco-budget, look for a local repair workshop or Mens Shed (where both men and women will be made welcome and given a helping hand). Here in Ennis, where I live, the local Men's Shed has come up trumps for many local projects; producing tree stakes for new community orchards and building bird boxes to help support wildlife.

Wood: untreated, treated and composite board

Natural untreated wood is the best option from a household waste minimisation perspective, because it can be used as kindling after all of its other uses have been explored. Avoid burning composite boards because of the air pollution generated by the glues and resins.

Larger recycling centres usually have a depot for waste wood. Ask what happens to the wood you bring to the centre. Recycling composite board back into chipboard is a more ecologically sound option than adding the glues and resins to the air by burning, or to the soil by chipping for landscaping. However eventually the wood will make its way to landfill or incineration, so best avoid it at the purchasing stage insofar as possible. This rules out most cheap furniture, but it is possible to have a more eclectic mix of furniture by looking in second-hand shops, antique shops etc. for interesting solid-wood furniture instead.

A pretty obvious replacement for fibreboard is hemp board. Hemp produces many times more fibre per acre than timber, and is a very versatile crop, with little or no requirements for pesticides. On a national level it's something well worth pursuing as an agricultural crop.

Remember that local timber from well-managed woodlands is an excellent, low-impact construction material. Roundwood construction requires a change in the design from milled timber, but it's no less effective and can give a very beautiful finish to a building or project.

If buying wood from a mainstream supplier choose Forest Stewardship Council (FSC) certified wood, whether natural or composite. For composites, look for plywood and chipboard with non-toxic binders and glues. Depending on the project I have in mind I sometimes opt for damaged or discarded wood at the hardware store. Occasionally I'll get it for free or at a reduced price, but either way it makes use of wood that might otherwise simply be thrown away.

Aggregates: gravel, sand, stone, leftover blocks

Sand and gravel do not have a large waste footprint, but minimise as always for energy and environmental reasons, to avoid the impacts and energy inputs needed to quarry, grade and transport them. Leftover sand can be used for children's sand pits. Recycled demolition aggregates are increasingly used to generate material for road construction while minimising the use of landfill space.

Builder's rubble can be used as driveway fill for easy on-site recycling of clean demolition wastes. When we started renovating an old cottage in West Cork we reused the demolition material (from the floors we dug out to damp-proof the building) as fill under a new patio.

Cement, plasters and admixtures

Keep different types of leftover materials separate so that each can be evaluated for reuse or recycling on its own. Keeping materials covered and appropriately stored ensures that they do not spoil. This maximises the percentage that can be used on-site, on other jobs, returned or recycled. If possible, use materials within their expiry date rather than storing until they need to be dumped. For products like cement, plasters, grout etc. they will start to go off from the moment they are opened. Either store them very carefully in an airtight container or give them away as soon as the job is done so that somebody else can use them.

Bear in mind that we can build perfectly good houses without cement plasters or concrete blocks at all. Eco-friendly materials such as cob, clay slip, wattle and daub, straw bale or timber all have a negligible carbon footprint compared to concrete. In our last house we had a porch that was built mostly of cob and plastered with a lime render. The structure was beautiful, complete with old coloured bottles built into the wall that glowed magnificently in the dawn sunlight. When it finally falls apart it will melt into the landscape and leave a few bottles that can simply be used again for the next building project, washed and refilled, or recycled; depending on the infrastructure available when the time comes.

Adhesives, fillers and fittings

For adhesives and related products, as with every other potential waste, avoid mixing. Adhesives and fillers are not easy materials to deal with since they often contain toxic or hazardous ingredients. They also have a limited shelf life, but are often used in small amounts. Minimise their use insofar as possible and buy only in the volumes that you need. Sugru is a colourful, mouldable adhesive that comes in small foil packets, so that for tiny jobs you only have the amount you need rather than a full tube of adhesive that will go off in your shed. This goes against the usual advice to buy in bulk to reduce packaging, but is valid if the alternative is wasting a large percentage of the product as well as the packaging itself. While the ingredients in adhesives are no ecological saints, some are safer than others. Also, fixing extends the life of a host of things around the house, keeping them out of landfill and avoiding new purchases.

Back to the more traditional fixings and fittings; nails, screws and fittings can be extracted carefully for reuse, or recycled for their metal if damaged.

Toxic and hazardous elements

Toxic and hazardous components can be present in insulation materials, wood preservatives, propellants, glues, admixtures, varnishes, paints and other commonly used builder's materials. All of these can be avoided – witness the fact that we've built houses for most of human history without them. However in modern life and modern construction, it's pretty difficult to avoid them all and still carry on life as usual. Minimise where you can. Substitute where possible using safer, more natural alternatives. If you find yourself opting for the standard off-the-shelf product, dispose of any leftovers as toxic and hazardous waste at your local bring centre.

Paints, varnishes, wood stains, preservatives, adhesives and other products and materials can be kept and used on other jobs. Sometimes there are paint collection points for community groups to utilise leftover paint, or list it on your local Freecycle network. This is a good way to ensure that leftover paint and other surplus materials get used. Plastic and metal waste generated by plumbers is generally clean and readily recyclable. Electric wiring should ideally be separated into copper and plastic before recycling the copper because of the potential for toxin generation.

When renovating our first home we tried to reduce the toxic and hazardous elements of any new materials that we bought for the job. We didn't always get it right, or have the time or energy to pursue our aims as much as we would have liked, but having the goal of zero waste in mind meant that the impact was much lower than if we had followed standard practice.

Finishes: tile, panel board, paint, wallpaper, varnishes, floor boards, vinyl

When it comes to finishes, usually the more environmentally-friendly options will have wastes that are easier to deal with than the standard products. For example it is safe to burn untreated waste timber, but toxic to burn waste lino/vinyl (PVC). High quality leftover materials can be reused on other construction jobs: timber, roofing materials, piping, aggregates, cement and admixtures, insulation materials, nails, screws, electrical and plumbing fittings and blocks.

'As new' leftover materials can sometimes be returned to the suppliers, cutting down on costs, waste generation and clutter. Otherwise look for a swap site in your neighbourhood to get full use out of the unused materials, or set one up if needed. Cardboard and plastics are often used as packing materials for deliveries to construction sites and can be recycled if they are kept clean. Burning rubbish on site is illegal and can be harmful to your health if plastics, composite wood products, paint or glues are included.

Tools

Don't buy low quality tools on impulse just because they are on special offer. Only buy what you will use and when you buy, get as high quality as you can afford. Bear in mind that metal tools are generally more durable and can be recycled more easily than plastic. Keep them in good condition to maximise their lifespan. Pass them on if you won't use them again, so they don't clog up your storage space. Doing this potentially allows somebody else to avoid making unnecessary new purchases.

Tool-sharing is a great way to have access to the tools you want without the cost, resource input or carbon footprint of a new purchase. Online resources and face to face tool libraries are growing in number all around the world. Have a look online for your nearest hub[91] to give idle tools a new sense of purpose or to get that last job on the list finally crossed off.

Travel and Transport

One of the joys of our era is the ability to travel wherever we want (if we live in certain selected countries, it must be said) and to enjoy the fruits of every climate available to us. Who before us in history, other than kings, queens and emperors, has had the selection of fruit and vegetables and meats and grains from the gardens of the world. Yet we have taken this to such an extreme that both Kerrygold and Anchor compete in the same global market for butter sales, from two opposite sides of the world.

The massive subsidies for dirty fuels mean that we essentially pay to support the madness of this international system,[92] rather than simply learning to love the things that grow best on our own doorstep. Waste, climate breakdown and wildlife loss are all byproducts of the process, as companies simply shift to the cheapest ingredient source without any need to account financially for the distances travelled or the processes involved.

Have a look at 'Chapter 4: Hidden Waste' for more on food miles. Meanwhile here are some ways to move towards zero waste as you travel.

Transport
In short, the closer to home your produce is grown or your products manufactured, the lower the energy input and waste. Moving things by air has a higher carbon footprint than ocean transport. If you grow it yourself your food miles and waste impacts will be as close to zero as possible.

Getting around
The car is a relatively invisible contributor to waste because it doesn't end up in your weekly bin. However the resources used to scrap and rebuild cars are vast, before ever you count in the maintenance of the roads and services needed to run a car. So if you use one, keep it well maintained and keep it for as long as you can. The combined resources required to maintain an old car are lower than those required to produce a new one, no matter how eco-friendly.

Walk, cycle or take the bus or train rather than driving where practical.

Telecommute rather than sitting in traffic, using one of the many online video conference packages on offer. As a waste issue on the train, a packed lunch will have less plastic than a wrapped muffin and a disposable coffee cup.

Car pooling services such as LiftShare, BlaBlaCar and GoCarShare can then get you from A to B while you're out and about – whether on your regular commute or abroad. These online services all help you find a spare seat on a journey rather than having to get in a taxi or even take public transport if you decide to live car-free, and are much more eco-friendly than driving alone.[93]

Keeping more of your journeys within walking and cycling distance is the best transport option of all. It is healthier by far, as well as costing less. At the end of the day, an old bike will be much less bulky to dump than an old car! Use your bike regularly to keep it (and you) fit. Keeping it well maintained will prevent it from rusting. A second hand bike is more environmentally-friendly than buying new and an older bike is also less likely to be stolen than a new one.

When you're out and about be sure to plan ahead to avoid the waste involved in something a simple as finding a drink of water. Bring your own water bottle, and a sandwich if you'll be away over lunch time and want to avoid plastic. Refill.ie and refill.org.uk provide maps of free places to top up your water bottle while on the move.

Holidays and travel

With online resources available now, reuse can be broader than just getting rid of our stuff to a local reuse group. Air BnB, Fair BnB, couch surfing, WarmShowers and BeWelcome are just some of the accommodation options available around the world that rely on a spare room rather than a new hotel.

In-home accommodation has a lower water and energy footprint than hotels,[94] with the added social benefit of meeting the locals. Other online resources include house swaps, work as rent, WWOOFing, skill swaps and other novel ways to travel. These can provide both affordable holidays and a way to really experience a new place more fully than simply dropping in as a tourist. Compare the many options available to find the one that suits you best.

Despite the hassle and carbon footprint of flying, I'm not entirely convinced that we'll wean ourselves off international travel any time soon. But it can change shape (more boat and rail travel, for example) and become more personal, experiential and eco-friendly if we approach it with a new perspective. One of my brothers made his way around the world in his twenties as a galley chef, living the international travel lifestyle without the carbon footprint or expense of flying. One way to reduce the carbon footprint of international travel is to tax aviation fuel (yes, it's currently tax free) and provide subsidies for train and bus fares. Read the advocacy section and start campaigning!

In the Office

Whether you manage day-to-day household finances or run a company, most of us have paperwork to deal with. Here are some zero waste ideas to help you along.

Paper and stationery

Whatever your business, whether consultancy, farming, accounting or bird-watching; you're likely to be the recipient of a lot of incoming post. This can constitute potential waste or potential resources depending on how you look at it.

In my office I use the reverse side of all suitable 'waste' paper for printing my own records and accounts and thus reuse incoming post and reduce the need to buy new paper. Standard A4 scrap paper can be reused directly if the reverse side is not specifically confidential. This is termed 'printer scrap', in our house and is de-stapled and stored alongside standard printer paper for printing drafts, lists and the like. For 'good' paper you can use 100% post consumer waste recycled paper to minimise the waste, energy and toxins in the supply chain. Also, by double siding all outgoing documents, you can reduce paper consumption by half.

Carefully opened envelopes can be reused by getting re-user labels from your favourite charity, or by printing your own logo on A5 stickers and writing the new postal address on that. Or simply cross out the old name and address and clearly write in the new one. If you have too many coming in to meet your own needs, ask a local charity if they would reuse good but used envelopes. It is a double bonus to reuse envelopes with windows. These are difficult to recycle due to the paper and plastic being glued together, and you don't even need a sticker to reuse them.

Pencils use fewer resources than pens and contribute less to landfill when finished with. Unpainted pencils can simply be burned as kindling when you get down to the stub. Good recycled stationary suppliers also provide a range of other eco-friendly and low-waste office supplies.

Another creative use for scrap paper is to use it to make your own handmade paper for use in craft projects. It can also be used as firelighters, wrapped around a candle stub or daubed in grill pan fat.

Incoming post and junk mail

After sifting out the useful paper and envelopes, all coloured or glossy paper should be sent for recycling rather than fire-lighting. Junk mail can be returned to sender (or at a poor second, sent for recycling). If the postal service is happy to offer a direct marketing service, I'm happy to write 'please return to sender' at their expense. My rationale is that if we all did that they'd have to raise the cost of mass marketing and make it less viable to distribute junk mail in the first place. If junk mail is a particular problem, make a small polite sticker for your door 'no unsolicited mail please'. It works most of the time. The happy side of GDPR is that it's trickier for companies to hang onto your address without your consent – making it that little bit easier to remove your name from direct marketing lists, one bit of junk mail at a time. The fewer marketing surveys you fill in or loyalty affiliations you have, the less junk mail will be generated on your behalf.

Storage facilities

Storage space is a constant issue in offices. De-clutter as much as possible and cut down on what you choose to store. Remember that every storage system, whether it be shelving, filing cabinets or box files, will need to be disposed of after the end of its useful life. Plan for durability and ideally for future separation, for reuse or recycling at the end of the product life. Recycled cardboard folders, binders, box files etc. are available from good recycled stationary suppliers. These are much easier to dispose of after their useful life is over than their plastic equivalents. If you already have the plastic ones, use them to the end of their natural life before you re-invest.

Increasingly businesses and individuals are using cloud services to back up communications and files. While this may save on space (and potentially on building an office extension) it carries its own waste footprint. Databanks require constant cooling. The electricity and air conditioning refrigerant gasses both have a very high carbon footprint. I'm not sure how best to approach this one because the rise in data storage seems to be ongoing. However if you practice minimalism in this area of life and declutter your electronic data as well as your physical stuff, at least you will reduce the amount of data simply dumped into data storage facilities in your name.

Computer/photocopier/printer/phone

Electronic items are typically large and bulky and are increasingly illegal to dispose of via the normal bin collection. Where computers, printers and copiers are necessary, plan for durability. The longer you get out of an item the better. Get things fixed where possible, rather than buying new ones. If you can postpone changing your printer for as long as possible, you keep it out of landfill or incineration, avoid the disposal cost and avoid the cost of a new one. Sometimes the difference between a 'friendly old printer' and a 'frustrating wreck on its last legs' can be a judiciously placed piece of card taped to the paper feed tray to keep it going smoothly!

Printer cartridge refills and recycling facilities are increasingly available. Ask at your usual stationery supplier or find a local charity that collects them for raising funds. We have been using a printer for years now that has refillable ink. We purchased the printer and an extra set of cartridges so we're never stuck. After that it's just ink that is needed, not extra ink cartridges. Even at that however, be judicious about what you send for printing so that your overall resource use is minimised. By printing at draft quality for internal documents you can save on ink.

Over 7bn smart phones were produced between 2007 and 2017.[95] The energy requirements, toxin loading and resource extraction required for these has been staggering. Phone companies rush to offer upgrades and new models, making the old ones obsolete obscenely rapidly. Buck the trend and keep your phone for as long as possible; buy second-hand to keep old but good phones in use rather than generating sales for new phones.

Workspaces

As human animals, one of the biggest impacts we have on the world is the size of the shells we carry on our backs – our homes, schools, offices and other work spaces. We can reduce this impact by shrinking the size of our buildings, cutting back on the need for new-build projects and doubling up on the use of existing structures. By using shared office space or working remotely from home we can free up space for other uses. Co-working can also be cheaper and more sociable than working in your own dedicated office space, as well as having a lower carbon footprint.

Chapter 7

Composting in Detail

Did I say it already? Compost is amazing! Composting takes all manner of waste food bits, garden trimmings and weeds and converts them into a seemingly magical substance called humus. This is part of the biological glue that holds a good friable soil crumb together, helping with both moisture retention and drainage. In the process, carbon dioxide is sequestered from the atmosphere into the soil in a form that enriches and enlivens it. Our soil is our future. That is becoming ever clearer with each passing year.

Hand in hand with composting is the whole area of no dig gardening and no till agriculture. By leaving soils intact, and using compost applications on the soil surface, we can increase the nutrient levels in the soil while at the same time providing the right environment for a group of fungi that produce glomalin. This sticky protein is produced by the fungal hyphae in healthy, undisturbed soils, and is even more important than humus in building soil and sequestering carbon (storing 27% of total soil carbon).[96]

We have lost almost a third of the world's arable land to erosion or pollution in just the past four decades.[97] The chemicals we add to our soils, along with ploughing and heavy machinery, all destroy the structure and life of the soil and the myriad of soil organisms within it. How we grow our food is threatening to destroy biodiversity, ruin global climate systems and ironically, completely undermine our capacity to feed ourselves.

However, here more than perhaps anywhere else we would do well to remember the words of the grandfather of permaculture, Bill Mollison: the problem is the solution. How we grow our food can become one of the main ways that we can build soil, sequester atmospheric carbon, reduce flooding and drought, have thriving waterways and an abundance of healthy fruit and veg.

There are many ways that agriculture can become part of the solution used to reduce the impacts of climate breakdown and to reverse pollution impacts and erosion.[98] Holistic management, silvopasture, regenerative agriculture, farm-scale permaculture, no till agriculture and biodynamics are just some of the terms that we are already hearing more of in conversations and articles about making positive change in the world. Composting, whether on our farms or in our back gardens, is an integral part of that world.

Adding compostables to the main rubbish bin is prohibited in many areas at this stage. A small kerbside composting bin is now a common sight on bin day. If your local authority composts kitchen and garden organics, then all is well and good. You could say that this chapter isn't for you. However if you embrace the rest of the information in this book you'll most likely eschew your weekly collection. That being so, unless you start composting yourself, you'll still have to have a bin collection service just for the privilege of having the makings of good rich soil taken away from your kerbside every week or two. Besides, there's all that rich garden soil to consider.

This is not a definitive book on household composting by any means, but I've presented some of the different basic types of composting to help direct you towards a method that might work well in your particular circumstances.

'Composting is easy' is the refrain from councils and environmental groups alike. However in my experience, it's actually not necessarily that easy at all... I don't seem to be very good at making good rich crumbly compost. Both my mother and my mother-in-law have far out-composted me on many occasions, despite my best intentions. Rather than treating this as a limitation, I'm going to celebrate being bad at compost making in the hopes that it encourages you to learn from my many mistakes. Remember that a heap of mixed organic material in the corner of the garden just takes time to heal most ills.

The main types of composting can be classified as follows:

○ thermophilic composting
○ cold composting
○ vermicomposting
○ trench composting
○ bokashi composting
○ plastic compost bins

Each method has its pros and cons. It is likely that no single system will fulfil all your requirements, but a combination of two or more systems may work perfectly. Try one or two that seem to suit your own life and lifestyle and be prepared to try something else the following year if the first choice does not produce the results you want.

Thermophilic Composting

Thermophilic composting is the standard aerated compost heap composting. This system of composting relies on the activity of heat and bacteria and other soil organisms to break down organic material into rich dark compost. The necessary ingredients are air, moisture and a good mix of 'green' and 'brown' organic material. This system needs some skill and care for it to work properly, as well as strength for the occasional turning. Rodents and flies may be a problem if food waste is included, unless you contain it effectively.

The site for the compost heap should be easily accessible and preferably located where it will get some sunshine, but shade from full summer heat. It should sit directly on the ground, to allow microorganisms and worms to enter and exit as they wish. The standard design for keeping a compost heap in a tidy shape is some sort of timber unit of about 1m cubed. This can be anything from three pallets tied together to form the back and sides, to a purpose-built stacking frame of timber squares. An open heap will suffice if you have the space and don't mind it being a little untidy. Stack new material around the sides first to keep the edge as upright as possible. Some jute stacking can keep it better insulated from the cold to help break down the outer edge material.

The compost heap is best made in one go, with 10-15cm thick layers of the different materials. Starting at the bottom layer, use sticks and twigs to allow air circulation. Next, put in some decomposed organic matter, such as rotted manure or old compost, to activate the heap. Follow this with layers of hedge trimmings, garden weeds, straw, paper, leaves and/or grass cuttings. Alternate the green and brown materials to avoid an excessive thickness of any one type of material. A sprinkle of lime is sometimes useful to control acidity if you are using a lot of pine needles or citrus peelings, but otherwise it may be unnecessary and can reduce the nutrients available in the finished compost.

Greens include nitrogen rich materials such as:

- grass clippings
- fleshy leaves such as nettles and comfrey
- young weeds and plants
- poultry litter
- manure

Browns include fibre rich materials such as:

- fallen leaves
- tough plant stalks
- cardboard, egg boxes
- crumpled newspaper
- thin woody prunings
- straw

There are intermediate materials that do not really qualify as either green or brown, but are perfect in the compost heap. These include food scraps, tea bags and coffee grounds, cut flowers, animal bedding from farm animals and herbivorous pets such as rabbits and tough green leaves such as rhubarb and bracken.

While the ideal is to layer the heap, in reality the materials don't always present themselves in nice clean bundles of just the right quantity. Flexibility is sometimes necessary, but bear in mind that a heap full of light twiggy branches will just start to dry out and a heap full of grass clippings will form a sludgy mess, so take care to layer it at least a little. If a glut of any particular material shows up on a busy day of gardening, set some of it to one side and then find another job to generate the corresponding green or brown material to balance it.

The 'green' material tends to be high in nitrogen, while the 'brown' tends to be higher in carbon. In reality compost heaps usually benefit from some additional nitrogen to get them heated up. Poultry manure provides a good rich burst of nitrogen, but isn't necessarily available in everybody's garden. Urine is much more readily available and is also a very effective nitrogen-rich activator. Apply at night, or out of sight,

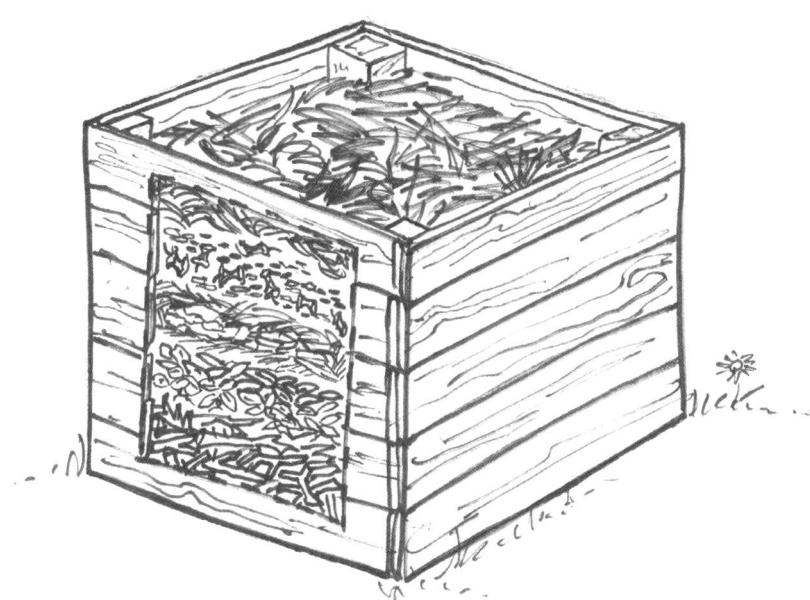

Compost material built up in layers of greens or browns, on a bed of twiggy material at the base for aeration and drainage. (The cut-away section in front is for illustration only!)

for minimum disturbance to your neighbours!

Having built the heap, water lightly and cover with a large piece of cardboard, jute sacking material or carpet (without the underlay, which tends to break apart after a while). Ensure that the heap does not dry out, as this will limit the growth of the appropriate microorganisms. The centre of the heap should generate a lot of heat, so check this every so often, using a pitch fork as a thermometer.

Once the heap begins to cool down, weeks or months depending on the ingredients, turn it once. This is laborious, but helps to mix the outer layers to the centre and to ensure a good kill-off of weeds and diseases. When this new heap has cooled, you should have rich dark soil. That's the theory. My reality is often a bit bitter than that, but it's still great for the garden. Work this finished compost into vegetable beds or use to mulch around fruit bushes or shrubs. It won't look like peat moss unless you sieve it, but this isn't necessary to get good growing results in the garden. Adherents of no dig gardening simply spread compost on the surface of their raised beds and let the worms do the hard work of digging it in over time.

Compost activators can be bought to kick start or speed up the process. I've generally avoided spending money on composting when it can be done for free, but recently I bought some QR (Quick Return) activator from Garden Organic. My usual cold, slow and somewhat underwhelming heap melted down into the rich brown crumbly stuff I'd always wanted. Granted there were still lots of twigs through it all, but sieving it would get those out if I wanted to.

You can also incorporate seaweed, either fresh or as a sprinkling of seaweed dust, to increase the minerals in the compost. Biochar can be added to the heap to provide a home for beneficial microorganisms. This is charcoal made from twigs, forestry thinnings or other low-value woody material, ideally mixed with microorganism-rich compost liquor, and used to enhance soil health. When the compost is used in the garden these bacteria and other microbes are safe from predators, and will be a presence for good in your garden. Biochar also aids both moisture retention and drainage and can fix carbon in the soil for hundreds or even thousands of years.

Thermophilic composting is a good method for making nutrient-rich compost if you have the space and commitment, but is not necessarily the most appropriate method for small gardens, for non-gardeners, or for food waste.

Cold Composting

If turning sounds like too much hard work then you can build a cold compost heap. This is essentially a heap of organic waste from the garden and kitchen, preferably layered, piled up to 1-1.2m high. This is topped with an inch of soil and a covered with jute sacking or an old wool carpet. The heap is then left to rot down into soil.

There a few drawbacks to this system as compared to thermophilic composting. It is a slower system, taking up to a year to break down. It is generally colder and more acidic, so lime may be needed in the garden to regulate the pH. It does not break down weed seeds or perennial roots. The fibrous elements won't break down as much, so it tends to be straw-like and rough. It may also attract flies if not adequately covered.

Despite these drawbacks, this can be a good method for those with enough space to leave the heap idle for a year and it has the advantage of being less work than most compost systems.

Seaweed, biochar or activators may all be added to a cold heap in exactly the same way as a thermophilic heap. Whether using thermophilic or cold heaps, the lack of rodent-proof containment means that meat, fish and cooked veg scraps should be avoided or you'll attract unwanted visitors to the garden. While rats and mice are a valuable part of the food web for barn owls and other wildlife, they are not always welcome guests in the garden. This is where some of the other methods of composting can be useful.

Vermicomposting

Worm Eat My Garbage was biologist Mary Appelhof's book[99] that introduced many people to the art and science of worm composting. Basically you build an environment in which the correct worms have the time of their lives: eating, breeding and making lots and lots of rich vermicompost.

Plenty sources of information exist to explain vermicomposting, step by step. The Irish Peatland Conservation Council website[100] espouses it as one of many alternatives to peat moss. Organic gardening books are another good source of information and can be found in any good local library.

The container is the first step with a worm bin. A converted rubbish bin will do fine, as will a purpose-built timber box system. Alternatively there are a number of purpose-designed systems on the market such as the 'Can O Worms'. Worm composting can be done indoors or outdoors, so can be adapted to apartment living.

Whatever design you adopt, the worm bin should be robust; well-drained and yet contained so that you can use the nutrient-rich liquid as a fertiliser; ventilated and yet sufficiently sealed to eliminate rodents and flies. If you are using it indoors, it should be sealed enough to keep in both worms and liquid. The worm bin should also be easy to use, both for filling and emptying. Timber boxes should be made of untreated wood.

First choose a bin, or make a box, with ample drainage. Cover the base with gravel to further facilitate drainage and then cover with a woven plastic sheet or jute sacking to prevent the compost mixing with the gravel. Place the 'bedding' material on this; a mixture of 50% shredded or crumpled newspaper and 50% compost, rotted manure or leaf mould. A handful of soil, crushed limestone or sand (or presumably sweepings from the kitchen floor) aids the digestion apparently.

A 'recycled worm farm' is espoused by Christchurch City Council in New Zealand.[101] This uses four old tyres placed one atop the other, sitting on a piece of corrugated metal sheeting. This system has the advantage of being well insulated for the winter. It also appears to be rat-proof, which is important, not least because the rats will eat your worms. For a large household you would need two or three stacks of tyres, because it is important not to let the uneaten food build up or it will rot rather than being made into worm castings.

Ideally you should keep the nutrient rich runoff as a liquid feed for tomatoes or other hungry plants, otherwise it will soak into the ground and may even cause pollution of the groundwater. If it does run off into the soil, then position the bin so that it can feed some tree roots in your garden and move it each time you empty it so that the nutrients don't accumulate unused in the soil.

I don't have any experience with indoor worm bins, so if you are trying this method be sure to do the research and background reading. *Worms Eat My Garbage* is a good place to start.

Whatever design you choose, site your worm bin out of direct sunshine and away from frost pockets or exposed windy areas. The worms thrive best at temperatures between 18 and 25°C. In winter the worm bin may need to be brought into a green house or shed if you want to keep the worms working optimally throughout the year (or just allow them to slow down a bit). I have seen worm composting systems working well in Irish conditions year round, with just a slight slow-down in winter. Sometimes however, the extremes of heat and cold have both killed off most of my worms at different times. In the recent hot summer I used a wet dust sheet to keep the bin cool, but moving it further out of the sun would have been better still.

Another factor when siting the worm bin is the location of the kitchen. The worm bin needs to be as easily accessible as possible, not a day's march through the garden, unless you would welcome the excuse to escape for a few minutes...

Next you need to find the worms, about 250-500g (or if you are counting, about 1,000 – plus or minus a few depending on their size of the worms).[102] The ones used for worm composting are those that thrive on rotting organic matter, so look first to a nearby horse manure heap or an existing worm bin or compost heap. Brandling or tiger worms (*Eisenia foetida*) are the usual species used. These are commonly sold for bait, so fishing shops are another port of call if neighbourhood supplies are limited. They are also sold by suppliers of worms and composting equipment.

Add the worms to the bedding material and cover with a good layer of newspaper to retain warmth and moisture and to keep the bin dark. Sometime the advice is to soak newspaper first, but my experience is that there is usually an abundance of moisture in a working bin as it is.

Now you are ready to begin adding the food. A converted rubbish bin will take most of the kitchen waste from about two people, so a bigger bin, more bins, or some additional type of compost system may be needed for larger households.

At feeding time, a number of pointers are helpful. A mixed diet is best for the worms, so include shredded newspaper and leafy garden waste occasionally. Add food only when the worms need it, so if you generate more than the worms can easily use, you may need a second bin.

Food scraps can be added in pockets in amongst the bedding material or as a layer of less than 5cm over the surface. The advice on what to include changes from list to list, but generally avoid anything non-organic in nature such as glass, metal, ceramic and the like. Seeding weeds will make gardening with the compost that bit weedier afterwards and diseased plants may carry those diseases into the next generation of plants in the garden, so a thermophilic heap may be better for these. Fish, meat, fat and dairy products are slow to break down and bought flowers can introduce a surprisingly high dose of herbicides and pesticides. However in modest quantities, these are all better put into compost than the bin. Dog and cat faeces can transmit diseases that can be dangerous to those handling compost and soil afterwards, so keep these in a separate system to avoid contaminating food crops.

Naturally the worms prefer being covered. Damp cardboard or newspaper is often recommended, but I find that this can be tricky to lift after a while as it begins to tear; it adds to the messiness and finickiness of the job. What works well is to place a layer of scrap paper such as a paper shopping bag or piece of old card in the bottom of a bowl or bin in the kitchen. Transfer the whole lot onto the top of the worm bin outside leaving the paper on top of the scraps. This seems to keep the worms covered enough for them to stay happy. Dig into the bin occasionally to check that the worms are keeping up to date with the food being added.

Maintenance includes feeding, checking and emptying. Worms can survive for four or five weeks without new food, so there is no need to stockpile before the holidays. Check the worm bin every couple of weeks to make sure the moisture balance is right. Water if it is too dry; add shredded paper or cardboard if it is too wet and address drainage. Check the worms too, that they are thriving and happy with the type of food being added, and the amount. There shouldn't be much of a smell, so odours indicate that too much food is being added and is beginning to rot. Reduce your food input to the bin if this is happening, to let the worms catch up. Surface area is the key to having enough capacity, so add another bin if the worms are overloaded. Checking with a small garden fork also aerates the bin, which helps to keep everything healthy. Too much citrus rind can lead to acidic conditions, so if this is suspected then stop applying them, and add crushed eggshells or crushed limestone for alkalinity.

Although the worms reduce the volume of waste considerably, emptying will be necessary at some stage. There are a number of ways to do it. If you have a low, flat box, press the contents to one side and add the new food at the other side. Over time the worms will migrate into the new food and the old compost can be removed. For larger bins the recommendation is to place their favourite food on the top layer for a week (presumably it takes a bit of observation to get to know them that well) and then remove this, complete with worms to buckets. Empty out the compost and then set up the worm bin again using the contents of the buckets as the new starting worms and bedding.

You can treat the resulting compost like concentrated fertiliser, both because of its high nutrient content and relatively limited volume. It can be used as a surface dressing for pot plants or patio planters; sprinkled into seed drills or mixed with garden compost. Some sources say it's unsuitable to use alone in pots and seed trays since it may be too rich for seed germination. If you collect the liquid from beneath the worm bin this can be used as a liquid fertiliser, diluted to the colour of weak tea before use.

For our own kitchen food waste we use two different worm bins. One is a wheelie bin turned into a wormery, with a tap fitted at the base to allow for containment of liquid and controlled removal of liquid fertiliser. The other is a Can-O-Worms unit, a three-tier stacking system of trays over a sealed base to catch and draw out liquid. The combination of the two systems gives us just enough surface area to keep pace with veg scraps from the kitchen. Sometimes we add additional scraps to the outside thermophilic compost heap, taking care to omit anything that may be of interest to rats.

Some people are reluctant to use a worm system on account of the worms themselves. If you are a bit squeamish, ask for help from a friend or neighbour, or buy an off-the-shelf, easy maintenance system, because they have a lot to offer in terms of reducing your kitchen waste and producing excellent compost in a completely rodent-free way.

Trench Composting

This is a fairly straightforward method of converting organic kitchen waste into soil. Dig a trench in the garden and add the vegetable peelings and other kitchen organics to it, one bucket at a time. After each application, cover with soil to prevent flies or odours being a nuisance. When spring arrives, plant courgettes, beans or other hungry plants over the trench. These will make good use of the nutrients generated by the decomposing material. The following year the trench will be ready for other crops. This is suitable for people who garden regularly, but even at that it can add a layer of complication to taking out the compost. Rodents can be a problem unless you earth up quite well or use Bokashi (see below), which makes the food waste unpalatable to them.

Trench composting is a good solution for dealing with occasional compost fiascos. If your chosen composting method didn't work as well as you had expected, then laying the resulting material in a trench and planting with hungry plants is an excellent way to recoup the nutrients and biomass and deal with the compost in a hygienic way. This method can also be used for pet-litter, but only in non-food growing areas.

Bokashi Composting

Effective Microorganisms (EM) are naturally occurring bacteria, fungi and yeasts that are sold in a balanced combination that can be used to enhance the health of soil, plants, water, humans and animals. EM functions in the compost in much the same way as yoghurt and probiotics do in the intestines, enhancing the action of the natural microorganisms. Bokashi is the name given to bran that has been mixed with EM solution for making compost.

With this form of composting the bokashi is sprinkled onto organic kitchen waste and then sealed into an airtight composting bucket. The food waste must be squashed down into the bucket to minimise aeration because the EM needs anaerobic conditions to thrive. Liquid generated by the mixture percolates through the bucket and is drained off at the base. This liquid is high in nutrients and beneficial microorganisms[103] and makes an excellent diluted liquid feed for the garden. It is also a good septic tank activator and drain degreaser, so it can also be poured down the sink neat if needed.

Rather than breaking down by the usual process of decomposition, the bokashi compost ferments. In doing so, the mixture becomes unpalatable to flies and rodents, so it can be buried in trenches in garden beds or can be added to a compost heap in layers without being as inviting to pests. By digging the bokashi treated waste into the garden you get the benefits of the nutrients from the waste as well as the benefits of the EM in the soil.

This type of system is best for gardens with enough growing space to dig in the treated material. Alternatively if you have gardening neighbours or an allotment, the treated waste is an active bonus for the soil – and the flowers and vegetables growing in it. Bokashi bran mix can be made up at home or can be purchased from specialist suppliers.

Plastic Compost Bins

There is a growing variety of plastic compost bins on the market. They are available from garden centres, garden suppliers and some local authorities. Even though they are generally either thermophilic or anaerobic in their action, they have been listed here as a separate entry because they are worth discussing on their own merits.

Plastic bins are generally used for kitchen waste rather than garden clippings and weeds, since the latter are usually too bulky. There are a number of different design types with corresponding differences in price and effectiveness. I have listed the different types as follows:

- freestanding compost bins/cones
- rotating or Tumbler bins
- green cone composters
- hotbins
- other options

The freestanding compost bins are the most commonly used plastic bins. These are usually conical in shape and sit either directly on the soil or on a perforated base. They allow worms in and let excess liquid out, while providing a barrier to rats. The older ones seem to have been modelled on coal scuttles, with a door at the bottom for removing finished compost, which actually acts as a handy, easy access rat-flap (which isn't ideal). My own experience with freestanding bins, with or without the perforated base, is that they have the potential to produce a relatively sludgy mess unless they are turned and aerated, which is tricky to do via a narrow top opening.

Nonetheless, my own occasional fiascos notwithstanding, these bins have the potential to be good for urban or rural gardens, being both tidy and inexpensive. The best results I have seen are where loading is modest and where rats are properly excluded, leaving the worms in peace to do an excellent job of turning the compost.

We've built a near zero compost system by fitting a tap to an old bin and raising it on some planks for ease of collecting compost liquor. Twigs at the base allow for drainage (you can cover with a square of groundcover sheeting if you wish, before adding the first kitchen scraps and worms). Some bricks or wooden blocks at the tap outlet help keep it clear. Keep adding food scraps and the worms will make their way up as they need to. Newspaper, cardboard or flour bags at the top will help the worms to burrow happily through the uppermost scraps without shying from the light each time the bin is opened.

Remember that rat poison can kill animals further up the food chain such as owls and hawks, so it's best avoided to protect wildlife. On a waste note, perhaps the best rat traps are the ones with the wooden base, so that when they finally break they can be disposed of easily and safely by recycling the metal and using the timber as kindling.

If you have a plastic bin that doesn't give the results you want, then use it for only a portion of your overall kitchen scraps. Use less rodent-tasty food such as citrus peels, onion skins and the like. The lower volumes of food will give the worms and microorganisms time to do their work and yield a better compost.

Rotating or Tumbler bins are freestanding or wall-mounted bins that can be rolled or rotated to aerate the contents. These tend to be fairly expensive but are rodent proof, report good results and are relatively easy to use. They tend to be good for small urban gardens, or anywhere where mess and rodents are particularly undesirable.

My brief experience with a rotating bin was that the food waste didn't ever really warm up enough to produce good thermophilic compost. The food got churned around and when I emptied the bin the results looked a bit like balls of elephant poo (I've never seen elephant poo, so I'm using my imagination here). I either added it to the garden heap or dug it into the main raised beds, but it wasn't nice crumbly compost by any means. That said, I've heard good reports of people getting nice rich compost from rotating bins as well, so perhaps trial and error is the way to proceed. The permaculture principle 'apply self-regulation and accept feedback' can apply in this case. From a waste perspective, if you change your mind on a compost bin, pass the bin on to someone else so it doesn't end up as waste and saves them buying a new plastic product.

The 'Green Cone' has a lattice basket base that is buried in the ground (as distinct from the standard freestanding compost bins/cones). These function as a disposal method for kitchen scraps rather than a way to get good compost per se. Bacteria and other soil organisms digest the food waste and the earthworms pull the compost out into the surrounding soil, fertilising the garden near the bin. These are a good solution for holiday homes with infrequent occupancy and a requirement for low maintenance. They also work well for small gardens where loads are small and food growing isn't a priority.

They are limited in the amount of compost that they process, and occasional emptying may be necessary to remove excess material. This material can be used as a layer in an aerobic compost heap or in a garden compost trench.

Because they are a disposal unit rather than a composting unit they can, according to their manufacturers, be used (to quote their literature) for fish, meat, poultry, bones, bread, pasta, soup, curry, fruit including peelings, vegetables including peelings, dairy produce, cooked food scraps, crushed eggshells, (paper-only) tea bags and animal excrement. If dog or cat faeces have been added, be sure to use any removed material with due caution.

HotBins seem to be a good mix of effectiveness and space saving. Getting a good hot heap going can be tricky for small volumes of kitchen wastes, so to answer this need the HotBin is an insulated unit that provides better conditions for thermophilic bacteria to get going. Alys Fowler, in her review in *The Guardian* gardening column,[104] describes it with abundant enthusiasm. In her experience, it swallows all manner of compostable ingredients plus a bulker in the form of woodchips or scrunched cardboard, and offers good rich compost in return; all in the tight tidy space afforded by urban gardens.

Other bins are coming on the market all the time now that composting has become so attractive. The design of the standard freestanding bins varies occasionally, producing greater rodent-proofing, different sizes and what-not. Before you choose a bin yourself, do a web-search or visit some friends and see what they have found to work.

At the end of the day remember that these are pretty bulky plastic items that will break eventually. If this is the surest way for you to get into composting then try them out. Try to be creative and see if you can compost with the resources already available for free in your garden or your local recycling centre. Experiment with the ideas presented and let your results guide you.

Composting conclusions

Remember that if the compost from any system is a bit anaerobic and smelly, you can bury the compost in a trench in the garden and grow courgettes or broad beans over it. Alternatively use cone or bin compost as a layer in the main garden compost heap, be that a thermophilic or anaerobic heap. If it's not an immediate success it will at least be unappetising to any rodents that may otherwise be tempted by food scraps.

If at first you don't succeed, remember that it is still more eco-friendly to try again than to throw in the trowel. Whatever compost you produce is sure to be better than the decomposition rates in a landfill. I remember hearing with fascination some years ago that a sample of excavated landfill material contained, among other exciting items, a pristine 25 year old lettuce,[105] preserved by the anaerobic conditions. Any attempt at home composting will be an improvement on that, and you'll be returning nutrients and biomass to the soil whether the system you use is excellent or sluggish. Of course with municipal composting in place, home composting isn't quite as crucial, but keeping all those nutrients and biomass in your own garden is still recommended.

Problem Ingredients

There is a list of items that are typically eschewed in composting advice. This list can include meat, fish and other bones, faeces from pets or people, oil, fat, cheese, bread, diseased plants, dairy, cereals, noxious weed roots or weeds in seed, newsprint, cardboard, ashes, used sanitary products and sawdust. If you are making rich humus for growing veggies, then there are plenty good reasons to keep some of the listed items out of your main compost system, but if your aim is zero waste then they still need to go somewhere and reincorporating them into the soil is still the best method to use. Here are some work arounds. Pick and choose which work best for you.

○ Meat, bones, bread, cheese, dairy or cereals can be fed to pets if they are still fresh enough to be healthy for them. Mix and match to ensure you get the right diet for the right pet.
○ Rotten food of any sort can be added to a sealed compost unit rather than an open heap. This will keep rats at bay while the worms and microbes work their magic. It will take longer to break down than standard ideal compost heap ingredients, but will get there in the end.
○ Fat offcuts, from fresh bacon for example, can be rendered down to make good lard. Healthy zero waste cooking fat and delicious crispy bacon bits as a byproduct! Less appetising offcuts or fatty leftovers can be wrapped in twists of newspaper to make firelighters. Old cooking oil likewise can be used to soak in newspaper as a firelighter.
○ Burning small amounts of meat and bones in a fireplace is a practical way to dispose of them, but it's illegal to cause a nuisance in the UK so be judicious about the amounts; and also illegal to burn non-listed fuels in smoke control areas in the UK. In the Republic of Ireland it's illegal to dispose of any wastes in a fireplace. (Perhaps best to get a good sealed composting unit instead.)

- Persistent weeds such as dock, ragwort, creeping thistle etc. can be piled up with weeds that have gone to seed and old briar roots or stems and left to decompose over a much longer time period than the main compost heap. You may not want to bring briar prickles back into your raised beds, but the finished soil will be ideal for a soil conditioner under bushes. You can also make a weed tea liquid fertiliser in an old bin. Pack a bin full of weeds and cuttings; top up with rainwater; cover securely with a lid for safety; wait for 6-12 months and then use to water tomatoes or other hungry crop, or as a general liquid fertiliser in the garden. Not necessarily as rich as comfrey tea, but a wonderful plant tonic nonetheless.

- Burning garden waste is illegal now in many areas, so this method of disposing of persistent perennial weeds, briars, trimmings and seeding weeds may not be available to you. Perhaps that's just as well. By composting garden biomass you can fix carbon in the soil in the form of rich humus, rather than releasing it back into the air. However if you have pyromaniacal tendencies, you may wish to dry the garden clippings and make your own biochar, which can be done in a mini kiln in your fireplace, and fixes carbon for hundreds or thousands of years.

- Coal ash should be avoided,[106] but wood ash will add potassium to your fruit trees. It's OK to add wood ash to the compost heap in small quantities, but if you have lots it may overwhelm the compost heap so best just scattered directly on the ground around trees and soft fruit. Fortunately the ash and the fruit tend to come at different times of year, making for cleaner fruit picking than would otherwise be the case.

- Newspapers and cardboard can be used as a weed-suppressing mulch beneath a layer of lawn clippings or woodchips. They are also useful to bulk up contained compost bins with dry material, helping to soak up the liquid that worm composting generates. Newsprint is now often oxygen-bleached rather than chlorine-bleached, making it safer to reintroduce into your garden; and the inks are generally soy-based and safer than the inks of the past.[107] If you're going to buy a lot of newspaper, like one a day, then it's worth contacting the media company and asking about the toxic load in their print-runs. Not everyone is happy to use the inks as part of their food chain, which is pretty fair, so do more research on this if you plan to use a lot.

- Sawdust carries a health warning based on the preservatives, glues and resins that can be used in furniture, pallets, treated wood or composite board. Look at the shavings on offer, and if they are clean real wood then the sawdust from the same supplier will likely be good too.

○ If you are working with clean real wood then it's fine for reusing as kindling or fuel. If you are using composites, pallets or treated wood, then perhaps not. If pallets carry the IPPC mark (International Plant Protection Convention) look for those that also say HT (Heat Treated) or KD (Kiln Dried). Avoid pallets with the MB code (Methyl Bromide, a broad spectrum pesticide which is classified as a highly acute toxin).[108] A fourth code DB, stands for Debarked. Unmarked pallets are generally used within countries where the international markings are not required. They're generally untreated, but depending on the source may have been exposed to spillages of toxins during transport.

○ Poo from pets and poo from you. Herbivore poo can go into the main compost heap, but faeces from carnivorous pets or people can still be composted. The reason we don't have mountains of poo sitting at the roadsides since the dawn of history is that it actually composts down pretty well, surprisingly enough. Dry toilets are far less polluting than flush toilets, so if you want to learn more read Joseph Jenkins' *Humanure Handbook*.[109] Cat and dog poo should be composted in a heap separate from your main garden pile. There are a load of nasties that you don't want to get into your bloodstream or body, so while you can compost away to your heart's content, it's important to do so with care. Review the pets section earlier in the book for more details. If you're using sawdust in a vermicomposting system for humanure or pet poo, make sure it's from clean, real wood.

○ Sanitary products can also be composted if you shop with care and buy only biodegradable products. Compost in a closed humanure composting system or vermicomposting system and then dig the compost into a dedicated part of the garden for further maturation. At our home we add these to the main humanure vermicompost system and then plant a crop of comfrey over a dedicated compost trench to recoup the biomass, nitrogen, phosphorous and potassium for use in the main garden compost heap. Comfrey leaves are very high in nutrients and this keeps any remaining pathogens or slow-to-break-down materials completely covered while allowing us to recycle the nutrients back to the garden via the harvested leaves.

Here are some that should stay out of your compost heap though:

○ Anything that is non-biodegradable in nature, such as glass, plastic, ceramic, masonry or metal. This includes the plastic windows in otherwise compostable paper envelopes and the polyester and elastane content in stretchy jeans or the elastic in underwear for example.

O   Anything that contains potentially toxic ingredients such as paints, solvents, batteries, household cleaners, conventional personal cleaning products and cosmetics. For that matter, you're better off not putting a lot of that stuff anywhere near you either.

O   Glossy paper is typically made with a filler/coater made of kaolin clay,[110] which can have low levels of radioactive elements within it.[111] The ingredients in the inks can be problematic too, given that glossy paper lends itself to high colour printing. It's also high quality, so probably best sent back for recycling, (which introduces those toxins into the recycled paper stream... so we're back to simplicity again, and not buying glossy paper in the first place.)

Lots of Compost? Get Growing

Once you have made compost, the next thing to do is to use it. All gardeners worth their salt will tell you how important compost is for getting good results. Gardening in some way, any way at all really, is the best method to utilise the compost that you generate from your kitchen and garden organics. What follows is for the non-gardeners out there who have more compost than they know what to do with. This won't be a problem for existing gardeners because the ratio of compost to veggie beds will always be the other way around...

If you find that you have lots of wonderful compost and no way to utilise it because your garden is modest in size, start small: a square foot garden is an ideal way to introduce children to gardening or to begin gardening yourself if you haven't done much of it before. It is also great for small urban gardens with minimum space. The square foot garden fits into an area 1.2m x 1.2m and is divided into 16 plots of one square foot each. Each plot is planted with a different type of vegetable or herb.

To start your own square foot garden, choose an area in a sunny position on relatively free-draining ground. The closer it is to the house the more you will get to tending it, and harvesting from it. Remember the Chinese phrase, the gardener's shadow is the best fertiliser.

Build a timber frame with four untreated boards of about 15cm x 3cm x 120cm. Two old scaffolding boards are ideal, each cut in two. You can paint them with raw linseed oil to preserve them for a bit longer if you wish. Lay some fairly thick overlapping layers of newspapers or cardboard on the ground to suppress existing grass and weeds. A full newspaper opened out flat is a good thickness to use. A layer of only one or two sheets will simply tear before you've even replaced the first shovel of earth.

A square-foot garden is an easy and productive way to start growing. With two old scaffold planks, some cardboard to cover the grass, and a bit of top soil and compost, you can be growing every type of veg you want in an afternoon. By using each of the 16 square foot sections you can have a large variety of plants in even a tiny back garden.

Plant a different vegetable, herb or flower in each plot, using the appropriate number of plants or seeds for each type. The number of each plant per bed will depend on their final size. For example, nine garlic bulbs, one potato, six cabbages (thinned to one or two as they grow), three broad beans and so on. As the plants mature, eat the thinnings. Surplus beetroot seedlings for example, can be eaten in salads, tops and all, while the remaining four can be left to mature. As each square foot plot is harvested in turn, add more compost as a surface mulch while planting the next crop of a different type of plant.

If you want to maximise the use of space, you can grow seedlings on in trays so that they have a head start by the time they go into the bed. Plant taller plants and climbers on the north edge of the bed, to avoid shading the rest of the plot. Harvest each crop as soon as it is ready and begin the next crop immediately. There are lots of good books available that give a description of what plants to place where for year round productivity. These are well worth reading if you want to get growing. Your local GIY (Grow-It-Yourself) group will also be happy to give advice and pointers if you are new to gardening. Look out for seed and plant swaps where you can get seeds, plants, pots, magazines, cuttings, herbs, advice and smiles to help you along.

If you live in an apartment you can still grow in pots or in a window box. You probably won't have enough space for all your compost though, so look for a neighbour, local allotment or community garden where compost would be welcomed. You may even find a community compost scheme nearby or set one up yourself to serve others in the same position.

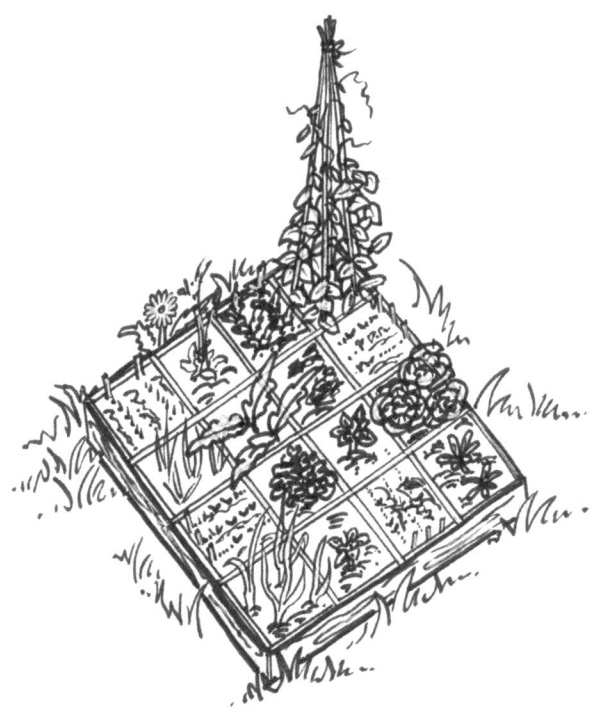

Place your frame over the mulch and fill it with a layer of compost on the bottom half and good loamy soil on top. Avoid moss peat since it contributes to the destruction of native peatland habitats which are important wildlife areas and a very valuable carbon sink for reducing climate breakdown. Coconut fibre isn't much better due to the carbon footprint in getting it half way around the world to your back garden. Mark out the 1.2m x 1.2m bed into 16 square foot plots with string, some tall willow canes, or locally grown bamboo. Now you are ready to plant.

Chapter 8
Zero Waste Culture

Is it possible to have a zero waste culture? Absolutely. Plastic is the clearest challenge in achieving zero waste, and it's a very new invention in cultural terms, so being plastic-free is completely within our grasp. However, given where we are now, achieving a zero waste culture poses many challenges.

In terms of the waste issue, each individual and even each country is located in a wider arena of production, consumption, travel, transport and transnational legislation. With plastic so ubiquitous and international trade so common, it is incredibly difficult to have a completely zero waste lifestyle or society. Nonetheless the potential for making positive changes towards zero waste is present in abundance, both in our homes and in wider society.

Remember that the globalisation of ideas and products can work both ways. Not only do products and services flow into our own towns and cities, but ideas about how to make positive changes do too. Whatever changes we make in our own house or our own town can be taken on social media out into the world to be an example to others working across the world on the same aims.

How do we best engage and help those changes along? If we are to aim for zero waste in our society, a multi-faceted approach is needed, covering every sector of society from production to consumption and back to production. We'll need people from every profession and with every skill to bring what they have to the table. Here are just some of the opportunities that present themselves:

Manufacturing

If you work at the manufacturing and design stage of a product, then design for reuse and easy repair, for recycling, for biodegradability. A big problem with reuse at present is that the design process for many products makes fixing things virtually impossible. With good design, the whole process immediately becomes more straightforward.

Keep the manufacturing process clean: eliminate toxic materials, substituting with safer examples where possible. This can be surprisingly easy. As an example some companies already use distilled water instead of toxic solvents in clean-room processes.

Many industrial wastes include an abundance of useful raw materials. This doesn't necessarily stop them being dumped, nor does it stop their classification as 'waste'. By keeping the different industrial waste types separate, and providing an incentive for minimisation, such wastes can be rerouted to other factories that need them. This can immediately reduce or eliminate the need for new mining or quarrying.

Design for full separation of the component parts for ease of repair, reuse or recycling. Given the abundance of single-use everything, we need to design for returnables and washables rather than for disposal. Washables such as hankies, nappies etc. are more sustainable than the paper alternative, because whole forests do not need to be destroyed to produce them and it reduces the plastic needed for every new wrapper to contain the disposable version. Half of the world's forests have been cleared for paper production, along with over 40% of current industrial wood harvest.[112] Between habitat destruction, energy consumption and pollution, you would be surprised how much havoc is wreaked just to get us paper![113] The figures are incomprehensibly large, and completely avoidable by simply choosing washables for tissues and nappies or recycled unbleached stationery products instead of the usual chlorine bleached virgin pulp version on the shop shelves.

By focusing on local sources for your products, you minimise the distance your goods have to travel. Stay close to raw materials, close to the consumer, close to access for special parts: ports, airports etc. and close to the returnables infrastructure and recycling processes. This spells smaller scale production and the development of a returnables and recycling industry within national borders rather than the present global market system. As we change policy and remove fossil fuel subsidies completely this will become the more economic option, as well as simply the saner, safer, and sustainable one.

Clearly this will suit some manufacturers more than others. Some products, like the tech industry, are by their nature more global in scope than others. Whichever area you are focused in, there are plenty opportunities to zero in on the things you have control over in your business and make positive changes. Not all changes are possible to make without shifts in consumer purchasing habits and government policy, so take the opportunity to stretch consumers with positive advertising campaigns, and lobby governments to make policy changes across the whole industry so that no one manufacturer loses out by taking the correct ecological course of action. At present, much of government policy is steered in the wrong direction by industry lobbying, so be a force for positive change to help move us all in the direction of cure rather than chaos.

Hand in hand with government policy, manufacturing and design are where much of the zero waste emphasis needs to start. Looking at microplastics for example, we know that the largest source of microplastics into the natural world, by a very wide margin, is from the wear and tear of tyres on our roads.[114] It is at the manufacturing stage that biodegradable alternatives need to be introduced. Washing of synthetic clothing is also a very large source of microplastics. While we as individuals can choose cotton or wool in preference to synthetics, to fit a microfibre filter to our washing machine or get a filter bag for synthetics in the wash, it is at the production stage that the real changes are needed, with a phase out of producing materials that pose such risks to marine life. Meanwhile local governments and water companies should fit microfibre filters on all sewage treatment plants as an interim measure. This won't be easy or palatable for manufacturers (nor for sewage treatment plant operators), but bear in mind that given the scale of the problems caused by microplastic pollution at the very bottom of the ocean food chain, such changes are vital.

It also makes good business sense. In this example and across the board, the companies that get in first with innovative products that are safe for use in the world will have a distinct advantage over the ones that drag their heels and delay taking a proactive stance.

Retail

In many ways all but the largest retailers can feel squeezed between government policy, suppliers and consumer demand. However, even so, there are so many opportunities to overhaul the way we sell to the public, even without government intervention and much needed policy changes. Many towns and cities have already set up zero waste or zero plastic shops where produce is supplied in paper, card, reusable glass or cloth wraps. It's very encouraging to watch health food shops, artisan food suppliers and farmers' markets making their own policies on plastics to cut it out of their supply chain.

For any retail type we can introduce return facilities for reuse of most items. Ink cartridges, shampoo bottles, milk bottles, veg boxes and the like are some examples. We want the contents, not the packaging. 'New for old' purchasing is growing for batteries and white goods such as washing machines and freezers to ensure safer control of hazardous elements. This means that refrigerant gases, for example, are more likely to be contained and reused rather than contributing to atmospheric greenhouse gases (at a rate of up to 17,000 times[115] the potency of CO_2).

Changes in marketing and labelling are also crucial. 'Recyclable' is a misleading term that looks too similar to the more eco-friendly 'recycled'. It should be unnecessary for any products to be sold that are not readily recyclable, so the term should be redundant anyway. Worse still, we need to stop putting 'recyclable' on packaging that simply isn't recyclable in the area where the product is sold. Clear and thorough labelling should include product miles and embodied waste, as well as the relevant recycling code for the different component parts of the product and the packaging.

If there was better product/packaging design to allow for easy sorting of recyclables it would make recycling much more straightforward and profitable. A composite waste labelling system like the current energy rating system would enable consumers to make a genuinely informed choice without needing a degree in environmental science. (Even with such a degree I find the area a minefield!) Repaired, reconditioned and reused goods would rate even better on the labelling scheme, because these would be effectively energy neutral.

Resource/Waste Management

In only a single generation we have seen whole industries move from a vibrant return and reuse infrastructure to a single-use and dispose model. Take your average pub, for example. I remember stacks of crates in the backyard of every pub, with empties all laid out for return to the suppliers. The deposit on crates and bottles usually cost more than the liquid contents. Now publicans not only have to pay outright for both the glass bottles that beer comes in, with no chance of recouping that cost again, but they also have to pay to have it hauled away again as a waste product for crushing and remanufacturing.

It's not difficult to have resource collection as an integral part of retail, as standard. Whether the containers used glass, plastic, card or whatever, having a seamless returns infrastructure will cut our waste generation and energy input dramatically.

Recycling is crucial as part of a zero waste society, but it cannot continue to be our go-to option. It has been used for too long as a way to ease the public conscience and pretend that it is a suitable alternative to proper reuse facilities. Nothing could be further from the truth. The energy involved in crushing glass and remaking a new bottle may well be lower than making glass from fresh materials, but still higher than return and reuse.

The infrastructure is already in place at local bring centres, particularly at the larger municipal collection facilities. With some minor changes in policy and the introduction of suitable buildings or good storage, these could become a hub for redirecting items from waste to reuse. This is easy to implement and can earn additional revenue for the council. Whether the resources are household items requiring repair, jam jars or compostables en route to rich soil, we need to start exploring ways to reuse the good and constructively recycle the rest.

As for non-recyclable wastes; really, is there any need to have something on the planet that simply cannot be recycled and needs to be put into storage in perpetuity, or result in resources incinerated unnecessarily? Personally I think not. In basic terms, if a product cannot be readily recycled into the same product again, or even downcycled into something of less value, but is still useful to society (and itself then recyclable) then is there a place for it at all?

Legislation and Government Policy

What policies can be introduced to encourage a zero waste society? What policies need to be changed to remove obstructions to that same goal? There are so many things that government can do to help usher in a zero waste society.

Taxation can be used to create a positive bias in shopping habits. The plastic bag tax is a good example of the immediate public response to environmental tax measures. Taxes can be used to tip the balance in favour of recycled goods: for example, to make recycled paper cheaper for the consumer, thus creating demand for the waste paper currently routed to landfill or incineration. Taxes can also help to balance for 'low waste' goods: for example cellulose insulation is easier to dispose of than polystyrene or fibreglass wool and as such could be made cheaper to help avoid landfill congestion at the disposal end of the lifecycle. Cellulose insulation also uses old newspapers, thus doubling the zero waste benefits.

Something to bear in mind is that the reason we're in the waste mess we're in at present is that plastic is lightweight, durable, versatile and ridiculously cheap. One of the reasons it's cheap is that many of the costs are borne by the rest of society and not by those who profit from the manufacture and sale of plastic. A price on oil and gas that reflects the true costs of climate breakdown and ecological damage would greatly help to redress this imbalance. The current scale of highly wasteful global trade in cheap products is only possibly by the huge taxpayer subsidies for the fossil fuel industry.[116] In recent years the modest increase in energy prices and carbon taxes has helped a little, but we're still way off the mark.

An effective way to step back from endless plastic waste (and associated climate breakdown and species extinction etc.) is simply to cap fossil fuel extraction at source. Instead of endless legal instruments to try and limit emissions from all the multiple industries that use energy, it would be much more straightforward to deal with the handful of companies that extract and sell oil and gas. The proposal from Irish think-tank Feasta (Foundation for the Economics of Sustainability) is called Cap and Share.[117]

Cap and Share takes the position that fossil energy reserves (and the biosphere that they damage) are a global commons, and thus their use should benefit everybody and not cause damage to the natural world on which we all depend. Rather than cutting the supply with no regard for the inevitable spiral on costs for everything that is made of plastic or uses energy in any way (that's a lot of knock-on price increases), Feasta's proposal is to distribute equal shares to everybody in the world; shares that the oil and gas industry buys back for the rights to extract the fuel resources. Thus those who use less than the global average will end up making a profit on the shares distributed to them. Those who use an average amount of energy (and embodied energy in the food and products they buy) will break even and those who use more than the global average will find their bills increased by the process. With electronic money movements and accounting the whole system could easily be facilitated.

Thus in one fell swoop, the extraction of coal, oil and gas would be limited to an agreed production output rather than endless policy measures to try to control emissions after the horse has bolted. The cost of the energy and waste associated with dirty fuels would rise, thus steering economic decisions towards zero energy and zero waste. Simultaneously those most vulnerable to energy shortages or cost increases would be protected by the distribution of share funds so that only those who consume energy and energy-intensive resources above the average would be out of pocket.

Another crucial policy change needed is to take money out of politics, or at least show a clear trail of the money movements behind policy decisions. With nearly one paid lobbyist for each of the 30,000 European Commission staff,[118] it is no wonder that so much of our legislation is influenced by industry. Imagine the good that could be done in politics if politicians were that heavily influenced by advocates for positive social and ecological change instead.

Perhaps we need to step up and be those social and ecological lobbyists ourselves. Here's a brief wish-list of policies that would help move us towards zero waste:

○ Introduce a deposit scheme on all packaging to maximise the recovery for reuse and recycling.
○ Introduce an additional levy on all non-reusable packaging to cover the costs needed to establish an indigenous returnables infrastructure.
○ Phase out packaging that cannot be easily and safely recycled.
○ Phase out single-use disposable items in favour of durable products.
○ Require greater product durability and repairability, particularly for electronic devices where the components require mining of semi-precious metals or inclusion of hazardous or toxic materials.
○ Introduce clear labelling on all packaging to show food miles, embodied energy and embodied waste.
○ Implement clear recycling labelling so that our current situation where

'not currently recyclable' is sometimes printed alongside the recycling symbol on the same packaging.

For food, there is a large trail of hazardous wastes involved in herbicide and pesticide use. All chemicals used in seed storage, ground preparation, crop growth, pre-harvest spraying and produce storage should be listed on the food labels so we can see what we're eating. This is a basic minimum responsibility to people who buy and eat the food, and would help to steer customer choices towards those foods with the lowest waste trail, which are also healthier in general.

○ Introduce Feasta's Cap and Share proposals to place a cap on global fossil fuel extraction and thus limit the wanton wastefulness of plastic production and disposal.

○ Discontinue the pursuit of economic growth as a measure of a nation's success.

The last two points are perhaps the most important and potentially effective measures of all. So, here's my proposal: when asked by governments in their public consultation processes (or when you just feel like writing a letter or email about the latest issue that jars on your conscience) close with something like this: "The issue that I have described in this letter, and many of the issues we face in the world today, stem from excessive fossil fuel use and the pursuit of limitless economic growth. I urge you to do all you can to introduce both Cap and Share and degrowth to address these twin issues..." If you want a nicer way to say degrowth, the French is décroissance – which I've heard has the advantage of sounding more like a delicious pastry than losing your job. Whatever the word used, losing your job shouldn't be necessary. There are generally more jobs to be found in a circular economy than the slash and burn economy we have at present.[119]

Local Government

In order to make the changes needed in the short timescale available we need to implement measures at every level of government. Local government often has the latitude to implement changes more quickly than at national or international level, so action here can rapidly yield positive results.

Local government can have a direct influence over decisions made around transport, energy generation and use, land-use and planning, waste, food and labelling, water quality and public education. To take some examples, central recycling facilities can become depots for surplus construction materials, electronics, furniture or fuel; local council buildings can implement immediate policies to purchase only recycled unbleached stationary; waste and ecological considerations can be written into the decision making processes, planning laws and regional policy.

On a wider level, funding and resources can be devoted to creating thriving rivers and streams, rewilding of parks, improvement of walking and cycling facilities. Supports can be put in place to encourage thriving farmers' markets, zero waste shops, repair shops and circular economy businesses in all their forms.

If we shifted the focus of local grants and incentives from economic development to ecological regeneration we would see a surge in ideas, solutions and paradoxically, new businesses to meet the growing demand for a more just, zero waste, regenerative world.

Money and the Economy

The root of all evil or the source of all good? Money seems to be what people want and what people blame in equal measures. As money moves around our society it seems to do untold damage. Pick any ecological issue that threatens our collective survival and you'll find a long trail of money that was needed to bring it into being. Mark Boyle, the Moneyless Man, argues that a gift economy would address many of the problems of our society. His book *The Moneyless Manifesto* provides wonderful examples of low-impact choices we can all make on a day-to-day basis.

However, moving to a moneyless society isn't necessarily something that is likely to happen any time soon; voluntarily at any rate (we are, inevitably, every bit as close to a wholescale societal and financial crash as we are to ecological collapse). I may be mistaken, but my bets are on something less radical – like moving beyond capitalism for example.[120] Our current economic structures of neoliberalism, or free market capitalism, essentially allow maximum freedoms to capital and minimum restrictions on how it can be used, with corresponding negative impacts on society and the natural world. Privatisation, deregulation, 'free trade' and austerity are the hallmark policies of neoliberalism. The past 10 years have shown us plenty examples of how these policies have wreaked havoc in our societies. The economic crash has hit hardest those who had least hand in causing it, whether through redundancies, homelessness, or those who are overworked yet chronically under-earning. In the wider world the impacts of ecocide, climate breakdown and mass extinction have continued unabated. It seems as if our political instruments have completely handed control of global governance to corporate interests.

What's very clear is that while large companies steer the ship, the challenges of climate breakdown, ocean warming, plastics in our seas, soils and food, biodiversity loss and soil erosion will continue. Companies usually have overt biases, , and often legal shareholder obligations to turn the maximum profit from what they do. If Naomi Klein's *This Changes Everything*[121] is correct, then neoliberal capitalism needs to be put back into its box and not be allowed to rule over government, media and society.

To take climate breakdown as one example, it should be a very straightforward matter to simply regulate the fossil fuel companies and introduce a cap on extraction. Just 10 companies control over 70% of the world's oil and over half the gas reserves.[122] Instead we are running around like headless chickens trying to control emissions from every conceivable human activity without ever thinking to turn off the tap that leads directly to those emissions in the first place. The only thing stopping carbon capping is the stranglehold that these incredibly rich and powerful companies have over governments, most media outlets and by extension the hearts and minds of the general public. Indeed, some of the top oil producers are owned by governments, which blurs the line further. But our enchantment with the economic process is waning. The rise of organisations like Extinction Rebellion and the school climate strikers belies the growing cracks in society's collective denial.

Whatever form of economy we adopt, we need to do it as a matter of urgency. As long as we worship at the alter of GDP (Gross Domestic Profit – the measure of the amount of money movement in an economy) we'll continue to make the choices that increase our return on investments, buy the cheaper products shipped in from goodness-knows-where, teach kids to pursue the high paying jobs rather than healthy lifestyles etc. In so many small ways we have all bought into the idea that economic growth is somehow necessary. It's becoming ever clearer to us that this beautiful blue bauble we live on is pretty small and that the pursuit of endless growth will destroy it.

In this light, GDP is an outdated and misleading index of societal health, at best. It counts hospital bills and prison numbers on the positive side of the balance, while low-cost, low-energy lifestyles are essentially negative since they don't increase the movement of money. More appropriate systems of measurement include Bhutan's Gross National Happiness (GNH), the Good Country Index and others.[123]

Hand in hand with this is the need to carefully examine the creation of money. About 97% of all money is created by private banks in the form of debt[124] (loans, mortgages etc.). The interest repayments require that a profit be made to repay the debt plus interest. Thus there is an ongoing requirement to generate ever more economic activity in order to keep this monetary agreement going. We have come to associate 'growth in our economy' with a healthy society, when in actual fact this endless requirement for growth is fuelling ecological destruction in so many different ways.

Alternatives include the circular economy, degrowth, co-operatives, the gift economy, economic democracy and socialising profits (spoiler alert: the current system socialises only the risks). These are not rose-tinted utopias. They are options that are available for us to investigate, select and push for with all our collective power. Because what is crystal clear is that we need a change for the better, sooner rather than later.

Meanwhile, what do we do with our own money and the trail of waste it is likely to leave behind it? Even if we watch how we fill our shopping baskets and apply practical simplicity to avoid the temptations of consumerism, there's still the issue of pensions, savings and loans to consider. Our pensions tend to pave the way towards a world we wouldn't want to retire into. A world of oil companies, weapons manufacturers, palm oil producers and the big seed tech companies that make both the genetically engineered seeds and the sprays that kill everything except the engineered plants. Time to switch to an ethical fund quick!

The same applies for savings. Credit unions and member-owned building societies help provide local capital for small community loans. Ethical banks such as Triodos and Charity Bank provide financial support for green and community-minded initiatives. These are easier to find in some areas than others, but more options are coming on stream all the time. There are also green micro-finance options that you can support, either as a philanthropic endeavour or as a green investment. Since these are newer and less regulated than conventional investments, you'll need to do careful research yourself to check both the green credentials and legitimacy of each project.

If you have a mortgage or loan – clear it as quickly as you possibly can. The earlier you repay your mortgage (literally 'death pledge') the more interest you'll avoid and the less you pay overall. If we're free of debt it frees us to step out of the rat race. Less spending, less waste, less running around doing stuff that we really don't want to be doing. From a waste perspective, remember that consumerism can be an addiction like any other. The happier you are the less you'll want to use your drug. The more time and space you have in your life, the happier you'll be. So if you borrow less and pay back quickly, then you'll have more freedom to get on with living the life you were born to live rather than simply paying back the men in suits.

Education and Media

I've lumped education and media in here together, because they're basically in the same game. They both help to steer public perception and priorities. At present both follow a predominantly economic focus. The main values and ideals are about getting a good job and gaining more material wealth.

Something that doesn't get much of a look in is the fact that the 6th mass extinction event on the planet is currently in full swing as a result of our modern way of life. Although humans are remarkably adaptable, our capacity to survive this extinction event is in question. The last event saw the demise of the dinosaurs, who had a much longer reign than we have had to date. The supreme tragedy is that this event is wholly avoidable. If we were to refrain from destroying ever more natural habitats, dumping plastics in the sea, putting poisonous chemicals on the land and pollution in the air, we'd address perhaps 90% of the causes of annihilation of the natural world. But this isn't something that our main sources of news and education pay much heed to at present.

What would happen if we were consciously to restructure our sources of information to create a bias towards pursuing collective solutions rather than perpetuating values that promote competition, money and status? Perhaps miracles could indeed unfold around us.

Imagine if the main focus of schools was to educate children to grow their own food, build their own home, generate their own electricity, sequester atmospheric carbon in trees and soils, start their own business doing something that they love and something that is kind to the world around them. Imagine if college taught solar installation, renewable energy generation and green building as standard; where the circular economy was a given rather than some radical notion; where the bottom line was the creation of healthy systems, rather than how much money could be extracted from every process. Every subject on the curriculum can be greened up to overtly support life on Earth rather than bolstering the consumer culture that is killing us.

Imagine if television, radio and newspapers delivered the news that actually matters. Would there be a revolution in the morning if we knew, collectively, that the oil industry experts knew about climate breakdown long before the rest of the world got their heads around it?[125] That public litter campaigns were first funded by the same industry that churns out plastic in the first place? As far back as the early 70s these campaigns sought to place the responsibility for litter onto the public, while the companies funding them actively blocked bottle deposit schemes and limits to plastic production volumes.[126] There are so many examples of businesses bankrolling research and policy to permit unhindered profits at the expense of our health and our living planet. Too often the media and government gloss over this completely, or until there is a public outcry to the point where they can't openly ignore the issue any longer.

Perhaps whenever we come up against the wall of public indifference and insurmountable quantities of soft plastic wrapping in our supermarkets we can take up our pen, keyboard or bodies and pursue advocacy as a tool for change in the schools and media that help steer public opinion in the first place.

Society and Community

We're all different. Thank heavens for that. This means that the solutions we adopt for societal change need to recognise that we all respond to life in different ways. As such, there are many different attitudes to the way we will manage our waste. Thus a different prompt or incentive is needed to encourage positive outcomes for each outlook.

At one end of the scale are those who just don't care where it is dumped. For them, legal enforcement is needed to prevent illegal dumping and burning. Tariffs, taxes and incentives will all help to steer purchasing decisions in the right direction without relying on altruism.

Then there are the NIMBYs, who just don't want a new landfill in their area, but are happy to send their rubbish off as usual to 'somewhere else'. Our contentment at exporting vast quantities of plastic waste off to Asia for recycling is a case in point. We now know that much of this has the potential to end up in the oceans of the world. However much we might be happy to dump our rubbish in somebody else's back yard, it still washes up on our shores.

As we grapple with the complexities of the problems and solutions, some will say "well, if the scientists insist that this, that or the other is safe, then surely it is". Alas, there have been too many examples of the scientific process being either suppressed or falsified to support the marketing process. Tobacco is the clearest example of this practice,[127] but over the decades many other industries have been implicated, including oil and gas, intensive farming, pharmaceuticals, genetic engineering, sugar, to name just a handful.[128] Not only do we need to have good scientific enquiry, we also need to have clarity as to the funding sources behind those studies.

This raises the question: should we all be self-sufficient smallholders? That is all well and good for those who choose it, but not an ideal lifestyle for everybody. We are interdependent beings and sometimes it's nice to simply fill a niche rather than having to grow everything from seed and do everything from scratch.

Beyond all these perspectives we need to realise that we have a major challenge around waste, energy and plastic pollution. There is a solution and yes, while readily achievable with full cooperation all round, it is complex and it will take work, dialogue and diligent action to achieve it.

A national or global zero waste strategy will have to be workable: many people will go to considerable lengths to pursue a more environmentally-friendly option, many more will not. Any strategy will need to work well to achieve practical reductions in the amount of plastic and other items that become wastes from being introduced into the marketplace in the first place. It will also need to be of some tangible benefit to people directly, whether that be via a financial incentive, a legal obligation or a social pressure, simply to bring on board those people who would not otherwise prioritise it.

A study carried out some years ago for Common Cause for Nature in the UK[129] concluded that most people are basically good. They want to see people and the planet flourish rather than simply pursue selfish motives. While government and business may appear to be laggards when it comes to making positive changes, bear in mind that when they were established, their general brief was to promote economic activity – as a measure of societal success. I believe we are at the stage in society where our collective desire to take care of the planet can now outweigh the momentum of old decisions that formed business and government in the first place. We're at a point where we are sufficiently aware and uncomfortable with our predicament that we can put enough pressure on businesses and governments to make the unpalatable changes needed for our collective survival and even our collective blossoming as a culture.

For the most part, people basically want to do the right thing; in business, politics, the media and at every level in society. It's time to get really clear that the right thing includes preserving life on Earth rather than letting it slip away through carelessness and neglect.

Under Your Own Roof

The household stage is arguably the easiest, because it is the area within which we have direct control. This has been the main subject of this book so I won't repeat it all here.

On a personal note, a quick calculation of people and resources in the world will show that we all have enough for our needs, but not for our greed. Many of our Western consumption habits fall into the greed category, unfortunately, because we purchase well in excess of any genuine needs that we have, at the direct expense of others. Here's a new slogan: 'Shrink your bin for world peace!' I suppose shrinking your shopping basket would be more accurate, but the end result will be the same. We can only expect peace in the world if we practice equitable consumption and fair international trading practices. These practices can start with ourselves and our own lives.

Zero waste does not require zero consumption. However our overwhelming desire for new and bigger and more will need to become a lot less extreme if zero waste and environmental sustainability are our goals.

Consumer Culture to Permaculture – Principle no.6

Another way to look at zero waste and creating change is with permaculture. Permaculture is a design application for creating sustainable systems that support human needs while protecting the environment. It was developed in Australia in the 1970s by Bill Mollison and David Holmgren, out of a growing awareness of the fuel shortages, food insecurity and wider ecological impacts that seem to be an intrinsic part of most aspects of modern life.

Originally conceived of for growing food, permaculture is now being used as a framework for examining all areas of life to create ecologically sustainable systems that can meet our needs without damaging the earth or one another.

Permaculture principles

One of the aspects of permaculture design is to optimise the beneficial relationships between the elements in any system to maximise the benefits to those managing the system – while at the same time protecting and enhancing the immediate and wider environment. It thus aims to design efficient systems that use the minimum of imported energy or resources to achieve the desired outcome. Permaculture principles are used as teaching tools and reminders during the design process.

In *Permaculture: A Designer's Manual*[130] and *Introduction to Permaculture*,[131] Bill Mollison and Reny Mia Slay outline an extensive list of permaculture principles to guide the process. A selection of these are set out in the Permaculture Association website,[132] as follows:

O   Relative location.
O   Each element performs many functions.
O   Each important function is supported by many elements.
O   Efficient energy planning: zone, sector and slope.
O   Using biological resources.
O   Cycling of energy, nutrients, resources.
O   Small-scale intensive systems; including plant stacking and time stacking.
O   Accelerating succession and evolution.
O   Diversity; including guilds.
O   Edge effects.
O   Attitudinal principles: everything works both ways, and permaculture is information and imagination-intensive.
O   Work with nature rather than against.
O   The problem is the solution.
O   Make the least change for the greatest possible effect.
O   The yield of a system is theoretically unlimited (or only limited by the imagination and information of the designer).
O   Everything gardens (or modifies its environment).

David Holmgren has distilled many of these guidelines along with his own ongoing observations and inherited wisdom to develop a set of 12 core principles,[133] set out as follows:

1. Observe and Interact
2. Catch and Store Energy
3. Obtain a Yield
4. Apply Self-regulation and Accept Feedback
5. Use and Value Renewable Resources and Services
6. Produce no Waste
7. Design from Patterns to Details
8. Integrate rather than Segregate
9. Use Small and Slow Solutions
10. Use and Value Diversity
11. Use Edges and Value the Marginal
12. Creatively Use and Respond to Change

Although many of these phrases are self explanatory, they only skim the surface of what permaculture has to offer for the design of any garden, farm, business or new national infrastructure plan. If you're new to the topic, see Maddy Harland's 'What is Permaculture?' series of articles in *Permaculture Magazine*.[134]

Permaculture ethics

"At the heart of permaculture", writes the late permaculture teacher and designer Patrick Whitefield,[76] "is a fundamental desire to do what we believe to be right, to be part of the solution rather than part of the problem, in other words a sense of ethics."

These permaculture ethics are summarised as 'Earth Care, People Care and Fair Shares'.

○ Earth Care acknowledges that we cannot live without due regard and care for the Earth's natural environment. Only with a sound ecological basis is life earth possible.

○ People Care states simply that you and I matter. A healthy environment is a necessary starting point, but we also need to develop systems that are socially just and support our growth and wellbeing.

○ Fair Shares recognises that there are limits to growth and consumption. There are many people and beings on the planet who all rely upon the available resources for their lives and livelihoods. There is plenty to meet all of our needs, but not to fill the greed for endless stuff and unlimited growth.

Principle no.6 in action

Produce no waste is a tall order for anybody living in modern western society. Yet only a few short generations ago, before the advent of cheap oil and the plastic and consumer items that go with it, waste was minimal and what existed was biodegradable. With the sheer volumes of waste generated today, we have become accustomed to a cultural norm that is directly at odds with our capacity to survive and thrive as a species (taking a lot of other species out in the process). Permaculture offers us a way to regain a way of being that doesn't cost the earth, while at the same time avoiding the drudgery that many associate with times gone by – and that many societies live with today due to the impacts and consumer habits of our wealthier nations.

From a zero waste perspective we can adopt all of the permaculture principles in our pursuits. We can *observe* what our current waste outputs are *and interact* by avoiding these at the purchasing stage. We can *catch and store energy* by installing PV cells or catching water from our roofs to reduce our waste impact of the wider electricity and water supply infrastructure. We can *obtain a yield* in many ways, whether it is through savings by buying in bulk, community resilience through interaction with our neighbours in creating community gardens, or spending time in nature for soul recharge rather than spending our time shopping as a pastime. The application of *self-regulation* is necessary for making the changes in shopping habits and by *accepting feedback* we can check in on our waste generation on a weekly or monthly basis and make adjustments to our shopping lists accordingly.

Renewable resources and services include a whole array of helpful allies such as gravity (to avoid pumps in our sewers and water supplies), trees (for shading a patio or providing fruit, nuts, coppice poles, planks or logs), sunshine (to warm well-designed buildings, heat water or provide power), or earthworms (to dig our gardens and farms for us while we sleep). None of these need *produce any waste*, and acknowledging them helps to keep us aligned with low-impact designs for the systems that serve us.

In *designing from patterns to details* we can look at a general overview of our waste sources before focusing in on how to address these one by one – or on a national scale, to provide a broad view of the infrastructure that is helpful in encouraging people to lead zero waste lives, and then focus on exactly how to put those measures into action. For example, first provide the deposit schemes and glass washing infrastructure, then push for public participation, rather than trying to get public buy-in for empty gestures. Following the same example, as we *integrate rather than segregate*, we can redesign all bottles and jars to a set of standard shapes and volumes so that all of the producers in our region can easily reuse containers from any source.

Slow and small solutions remind us that walking or cycling to a farmers' market for locally produced food has many benefits over high-cost, high-impact imports and exotics. Modest house sizes have modest embodied energy and waste inputs and are easier to heat and maintain. By *using and valuing diversity* we can enjoy learning how to cook the veg that is in season as the year turns. Whether on a garden or farm scale, crop rotations and a diversity of livestock breeds and vegetable seeds all offer resilience to shifting climatic norms. In the reminder to *use edges and value the marginal*, I think of autumn blackberry picking. From a zero waste perspective this principle puts me in mind of innovative reuse of waste streams by one industry as a raw material for another.

Life shifts and changes all the time. We can *creatively use and respond to these changes* in many ways: by recognising gluts at our farmers' market and taking the opportunity to make relish or jam; by engaging in awareness or advocacy in response to high-publicity events in the media; or by simply recognising when it's time to rest and recharge after a lot of time on a project or campaign.

As you can see, many of the principles are the basis for elegant design anywhere and are already a way of life for many indigenous cultures around the world. They can all relate directly back to the principle no.6: *produce no waste*. What permaculture offers is a conscious design process whereby we can recognise (literally to re-know) the beneficial relationships in nature that can help us meet our needs in a sustainable and even regenerative way. With permaculture we can shift from our consumer culture to a more durable, permanent way of life.

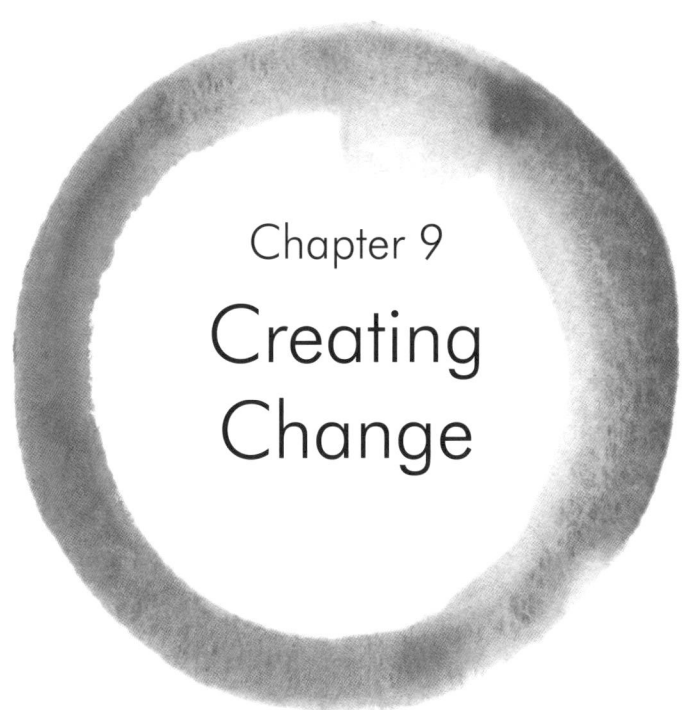

Chapter 9

Creating Change

Hand in hand with working to shrink our own bin and our own shopping basket we can also have a valuable influence on the wider world. Whether we know it or not, we contribute directly to the choices offered on shop shelves, to what manufacturers produce in the first instance and what governments incentivise or discourage through taxation. Whenever we shop, whenever we go to the polling booths, whenever we even talk about the state of the world: we're putting our money, votes and energy into shaping the world that's just around the corner.

But then why don't things ever change? Well, they do. All the time. Just rather more slowly than many of us would like, and sometimes in a direction that's less than reassuring. However, we can take this process to a higher level and actively engage in advocacy to herald in the world we want rather than leaving it to chance.

Examples of Advocacy in Action

As tempting as it is to feel disheartened and dejected by the seemingly slow pace of change, let's remember a few things. In the time of Henry VIII, homosexuality was punishable by death. By the time of Oscar Wilde this was reduced to a term of imprisonment. Now, through the concerted awareness raising of many people, gay marriages can be held in many countries.

Women couldn't vote on these islands a little over 100 years ago.[135] Not all men could vote until that time either.[136] Similarly in the US, the right to vote came first to wealthy white men and moved slowly towards being ever more progressive between 1776 and the present day.[137]

Slavery was once the mainstay of the global economy. Blacks in the US couldn't sit with whites on public transport, go to school together, or work together. Legally-enforced segregation was abolished only as late as 1964.[138] Queensland Aboriginals couldn't own land until 1975.[139] The list could go on in country after country and example after example.

All these things changed for the better only through a concerted process of putting pressure on governments to do the right thing instead of the old thing. If nothing else, this shows that we grow in our capacity for compassion and that advocacy can be effective.

Here in Ireland where I write, we've had a recent string of examples of advocacy paying off. Smoking is now banned in public buildings, making the habit far less appealing to our young people. The tax on plastic bags has played a major part in cleaning up our countryside. Remember what our road verges and hedgerows looked like before the tax was introduced? Onshore fracking was banned in 2017.[140] The state is also en route to become the first country to divest from fossil fuel shares and investments.[141] In a nutshell, advocacy has been very effective in many of our government policies.

On the path towards zero waste there have been victories all around the world. In exploring ways to reduce ocean plastic, Natalie Fee of City to Sea was contemplating the merits of a broad publicity campaign to educate people about flushing plastics down the loo. Instead she opted for the nine supermarket sustainability managers; a much easier focus than the entire population of the UK.[49] As a result of her advocacy work, plastic cotton buds are now being completely phased out of the supply chain in the major UK supermarkets.[142] One step closer to plastic-free seas.

But do we really need to engage in advocacy? Surely if we just make changes in our own lives, things will change for the better anyway. Well, to effect the changes we need to see in the world, personal change is essential, combined with joining together to work for broader change as well. Part of that work is bringing problems to the attention to those who can make decisions, and holding them accountable for their actions.

It's all well and good using a compost toilet to reuse biomass and nutrients, but if compost toilets aren't in the national codes and standards, their uptake will be limited and we'll continue to see septic tank pollution and fertiliser runoff from our fields (two sides of the same coin). By all means, holiday at home rather than flying, but as long as aviation fuel is classed as tax-free, there will be a taxpayer funded incentive to go to Majorca instead of Mayo or Mull on holidays.

Our incentivising of fossil fuels goes a lot further than aviation fuel though. Vast amounts of tax revenue subsidise some of the richest industries on Earth and help fund the main driver of climate breakdown, the fossil fuel industry, to the tune of some $70bn/yr; many times that of support for renewable energy, at under $20bn/yr.[143] Driving less is certainly important as an individual choice – but that our hard earned tax money should be paid out in subsidies for oil and coal and gas and peat is simply beyond belief. Surely there's a petition for that! (Oh, there is: www.stopfundingfossils.org, backed by over 60 petition sites and campaign groups from around the world.)

Perhaps the most recent examples of advocacy in action come from the Extinction Rebellion protests and the school climate strikers. In less than a year from their initial commencement both Britain and Ireland have both declared climate and biodiversity emergencies. This is more than simply a symbolic victory. It shows the immense power that ordinary people have we they simply take to the streets and disrupt the status quo in a co-ordinated, non-violent way. It also brings the much needed conversations about climate breakdown and species extinctions into the mainstream media. The time is ripe for starting or joining with a local group and getting your feet on the ground.

Where to Start?

Advocacy comes in many forms. Letter writing; emails; phone calls; online or paper petitions; marching in the streets; publicity events to highlight an issue that might otherwise be quietly swept under the carpet; non-violent direct action. All these are forms of advocacy. With all these actions, we're actively stepping up to the line and saying: "change is needed, we call on those responsible to take positive action".

In all likelihood you'll act only on those issues that frustrate or scare you most, but be polite. The Common Cause Foundation surveyed thousands of people across the UK for their report Perceptions Matter.[144] They discovered that regardless of age, political persuasion, gender or region, about 74% of respondents hold values that are essentially compassionate rather than selfish. At the same time, 77% of people assumed that other people generally hold selfish values rather than the opposite. In a nutshell most of us are basically compassionate, but believe that others are basically selfish. So, however frustrating policies and practices may seem, the person you're reaching out to is more likely to be compassionate than not.

What this means is that whatever your belief around the nature of the politicians, CEOs or anybody else you might be writing to; chances are that most of them are essentially good people wanting the best for themselves, their families and the world around them. If we hold that in mind during our interactions, it can help us to propose positive solutions to the challenges we're addressing, rather than venting our anger on the limitations of current policies or business practices. Also, it's a lot easier to say yes to somebody who is being considerate, than to somebody who is being irate and impatient. Changing the world will take all of us: CEOs and politicians included. Let's cultivate an attitude of collaboration rather than division to bring about the changes that are needed.

I'm not saying there won't be differences in opinions, values and priorities. Merely that if we can see the common humanity in one another as we work towards our goals we're more likely to build an equitable, peaceful world than if we alienate and blame one another.

Taking Local Action

Hand in hand with advocacy there are many practical actions we can take to engage our communities, to educate local businesses, councils and decision makers, and to clean up our small patch of the world.

Forming a local group

Many zero waste groups are popping up around the world, born of frustration at government heel-dragging and a rising sense of panic at the direction that we are taking. It's pretty easy to print off a few posters (on scrap paper) and advertise a meeting, or to share a post on social media inviting people to join you in a local campaign. We recently held a series of zero waste workshops in my town and here are some words of wisdom from our guest speakers and participants:

○ Take care not to overcommit and burn out. Sure, there's lots to do, but there are 7 billion of us around to do it. Just take on what you can commit to, and do that well. Don't worry about the rest. (Seriously, we need to put fear to one side because there's too much work to be done to let it hold us back.)

○ Take on manageable, practical projects so that the results of your time input will be visible. This helps keep the morale up and also helps to get others on board. The more successful something looks, the more people want to join in.

○ If you can measure your results, then you can advertise them more easily to better spread the message. One local campaign near where I live, Make Kilkee Plastic Free, diligently measured the amount of plastic removed from the supply chain by working with local shops. Rather than simply saying, "we removed a lot of plastic" they can confidently state "we estimate removal of over 400,000 items of plastic from the waste stream in Kilkee in 2018."[145] This is much more impressive and spurs people on to do more.

○ Don't let cynicism and negativity get the better of you or your group. It tends to sap the energy needed for the next step.

○ Celebrate small victories! The scale of the challenge can be over-whelming, but remember that each small local step that you take is being mirrored by so many other communities around the world at the same time, all working in a similar direction.

Local clean-ups

All across the world there are groups that are quietly and diligently picking up the mess made by big plastics manufacturers. For as long as I can remember my mother has picked up litter along the strand where we grew up. The beach wasn't always clean, but if she missed a week it became quite noticeable. She's by no means the only one. The Big Beach Clean organised by An Taisce (The National Trust for Ireland) and the Great British Beach Clean have events all around our coasts every year.

For those without easy access to a beach clean, there is the National Spring Clean in Ireland and the Great British Spring Clean which both rally around for volunteers each year for a sociable deep clean of litter in towns and villages all over. Increasingly we are coming to realise this isn't just about unsightly litter. It gets into streams and rivers and ultimately into the oceans and the marine food chain.

For greatest impact, bring along old sacks (discarded fertiliser bags are perfect) and separate out clean materials for recycling. Often the local council will offer to pick up collected rubbish if you phone them and arrange it in advance. If you separate out recyclables you may need to bring these to the bring centre yourself.

Reusing rubbish

One way that the global waste mountain is being addressed is by reusing rubbish as a raw material. On a large scale, tyres are being broken down as a component in road construction[146] and reuse of suitable demolition wastes is already written into European law.[147]

On an individual or community level eco-bricks can be made as a building materials. You simply stuff clean PET bottles with clean, dry soft plastics until filled to the gills, and cap them. Gather about 8,000 of these if you want to build a house,[148] or less if it's just for a shed. Either way it's a lot of time, but at no cost, and in a way that mops up plastic rubbish in the process.

A similar approach is the earth-filled tyre construction used for building Earthships[149] The filled tyres are strong, durable and have good heat-retention properties to store the warmth of the sun through the night. Earthship dwellings are designed to be off-grid, autonomous buildings that offer water harvesting, sewage treatment, energy generation, temperature regulation and food production.[150] Not a bad line-up for any building design.

A more straightforward way to reuse rubbish is to source your building materials from your local recycling centre, if that's an option where you live. Local builders, glazers, carpenters and builders providers are also potential sources of useful waste materials.

However you choose to create change, remember that your actions are part of a worldwide movement towards a healthier planet. Whatever actions or advocacy you choose to engage in, keep a positive outlook on life. Some letters and campaigns will work. Some won't. Some local initiatives will take off and bring about unexpected connections and excitement. Others may be a bit of a chore. That's the nature of it. Expect the best from other people and from life. I like to think that this expectation alone will help to bring it into being. Whether I'm right or not I don't know, but it makes me feel better and helps me to do the good work in front of me.

Chapter 10
Hold the Vision

There are challenges before us, but our capacity for imagination, creativity and love is present in such abundance that really we should have nothing to fear. Actively look for the signs of positive change, the signs of a collective realisation that we are making the shift that is needed.

Something I find incredibly uplifting to watch is how so many countries around the world are stepping forward to address aspects of this positive change. France has voted to ban the dumping of food from supermarkets that is past its sell-by date but still good to eat, in favour of distributing it to those who need it.[151] In 1994 Denmark was the first country to charge a tax (at import and manufacturing stage) on plastic bags and encourage shops to pass this directly to customers.[152] In 2002 Bangladesh went a step further and banned them outright.[153] Such taxes or bans have since been followed by a host of countries from every continent.[154]

China has deployed its army to plant millions of trees to help combat climate breakdown,[155] while volunteers in India set a world record of 50 million trees planted in one day.[156] Bhutan is now carbon negative, through a combination of tree planting and other environmental measures.[157] New Zealand has declared that land and rivers are entitled to a voice of their own in law.[158] Bolivia has granted Mother Earth equal status to humans under law.[159] Wales has a new building code, One Planet Development, that allows for low-impact dwellings to be built on land being used in an ecological manner,[160] which allows people on modest incomes to afford a low impact life on the land if they so choose.

Examples abound. The ones listed here are just a snippet of the changes unfolding around us. Humanity from every culture and corner of the world is looking at the madness of our era and wanting to change course for the better. Certainly old habits die hard, and changes can be slow in coming to fruition, but the changes are coming steadily.

The aim of this book is to encourage and assist you in your move towards zero waste, and in so doing to minimise your environmental impact. Having taken the time to read through these pages, take the time to put it into practise, taking it at the pace that suits you.

Have fun with the process. When the book has served its useful function for you, pass it on, sell it or donate it to a charity/thrift shop, but don't let it hang around cluttering up your bookshelves, taking up space...

Embrace advocacy in large and little ways. Whether it's a conversation with a neighbour, a request in a local shop, a letter to a government department or CEO, or an act of non-violent civil disobedience. Practising voluntary simplicity and making changes in our individual purchasing habits are invaluable as a way to help us live with integrity, but the course of society shifts even more when we bring these changes onto the wider stage.

All the while remember to hold onto a positive vision. Not denial of the problems, but an awareness of the potential we have for improvements and regeneration. We live on a beautiful planet. Not only that, but we are a wonderful species. Our capacities for joy and love, for ingenuity and fun, for care towards one another and our planet are present in abundance. Imagine living in a world where these are our guiding values. That's a world we can move towards with each conscious breath we take. Let's do it together now.

Resource Links

The whole area of zero waste has grown and expanded in the past decade. The plastic-free options for kitchenware, bathroom necessities and clothing are greater than it has been in a generation or so. (Of course, not too many generations ago, plastic hadn't been invented yet.) To meet this growing desire for low-impact lifestyles, more and more suppliers, books and online resources are becoming available all the time.

Rather than taking up space here outlining a list of resources that would rapidly go beyond their best before date, I've gathered some useful references and resources on: www.wetlandsystems.ie/wastetips.html

The website provides links to useful sites, publications or suppliers that I have found to be helpful in writing *Towards Zero Waste* and in reducing our own household's rubbish. There are many, many more sources of excellent information on waste, environment, composting, gardening and other related subjects that are not listed, so follow up any areas of interest further via your local library, advocacy groups or online if you cannot find what you are looking for on the website.

www.wetlandsystems.ie/wastetips.html

References

1. IPCC (2018) *Special Report – Global Warming of 1.5°C*. Intergovernmental Panel on Climate Change. www.ipcc.ch/sr15

2. Damien Carrington, 6 Sept 2017, 'Plastic fibres found in tap water around the world, study reveals'. *The Guardian*. www.theguardian.com/environment/2017/sep/06/plastic-fibres-found-tap-water-around-world-study-reveals

3. David Strahan, 1 Jun 2016, 'The Oracle of Oil: the man who predicted peak oil' *New Scientist*. www.newscientist.com/article/mg23030762-700-the-man-who-predicted-peak-oil

4. Manon Flausch, 22 June 2017, 'EU recognises bisphenol A as an endocrine disruptor'. *Euractiv*. www.euractiv.com/section/endocrine-disruptors/news/eu-recognises-bisphenol-a-as-an-endocrine-disruptor

5. Caleb Hellerman, 16 Jan 2010, 'Limit infants' use of Bisphenol-A product, FDA says'. *CNN*. http://edition.cnn.com/2010/HEALTH/01/15/fda.chemical.bpa/index.html

6. David Biello, 30 March 2010, 'BPA a "chemical of concern" EPA makes it official'. *Scientific American*. https://blogs.scientificamerican.com/observations/bpa-a-chemical-of-concern-epa-makes-it-official

7. Joseph Mercola (2015) 'BPA is fine if you ignore most studies for it'. *Mercola* https://articles.mercola.com/sites/articles/archive/2015/03/25/health-risks-bpa.aspx

8. Cancer Research UK, 4 Feb 2015, '1 in 2 people in the UK will get cancer.' *Cancer Research UK*. www.cancerresearchuk.org/about-us/cancer-news/press-release/2015-02-04-1-in-2-people-in-the-uk-will-get-cancer

9. SIPRI, 29 April 2019, 'World military expenditure grows to $1.8 trillion in 2018'. *Stockholm International Peace Research Institute*. www.sipri.org/media/press-release/2019/world-military-expenditure-grows-18-trillion-2018

10. GPI, website visited 17 May 2019, 'Glass Recycling Facts'. *Glass Packaging Institute*. www.gpi.org/recycling/glass-recycling-facts

11. Boris Borchev, 26 April, 2016, 'The future of the glass bottles – refilling or recycling'. *CASI*. www.casi2020.eu/blog/posts/the-future-of-the-glass-bottles-refilling-or-recycling

12. BBC News, 20 Feb 2014, 'Computer recycling West Africa style'. *BBC News*. www.youtube.com/watch?v=-JXDrlvShZKU

13. Ole Ryborg, 16 Apr 1997, 'Denmark battles to keep its ban on canned-drinks' *Politico*. www.politico.eu/article/denmark-battles-to-keep-its-ban-on-canned-drinks

14. Pirog R.S. and Benjamin A. (2005) *Calculating Food Miles for a Multiple Ingredient Food Product*. Iowa State University Digital Repository, USA.

15. Metropolitan Transport Research Unit (2014) *Heavy goods vehicles – do they pay for the damage they cause?* MTRU, UK.

16. Hann S., Sherrington C., Jamieson O., Hickman M., Kershaw P., Bapasola A. and Cole G. (2018)

Investigating options for reducing releases in the aquatic environment of microplastics emitted by (but not intentionally added in) products – Final report. Eunomia, Bristol, UK.

17. Project Drawdown (2014-2019) 'Materials – Refrigerant management'. *Drawdown*.

18. Adam B. Smith, 8 Jan 2017, '2017 US billion-dollar weather and climate disasters: a historic year in context'. *National Oceanic and Atmospheric Administration*. www.climate.gov/news- features/blogs/beyond-data/2017-us-billion-dollar-weather-and-climate-disasters-historic-year

19. Alice Facchini and Sandra Laville, 17 May 2018, 'Chilean villagers claim British appetite for avocados is draining region dry'. *The Guardian*. www.theguardian.com/environment/2018/may/17/chilean-villagers-claim-british-appetite-for-avocados-is-draining-region-dry

20. The World Bank, 20 Sept 2018, 'Global waste to grow by 70 percent by 2050 unless urgent action is taken: World Bank Report'. *The World Bank*. www.worldbank.org/en/news/press-release/2018/09/20/global-waste-to-grow-by-70-percent-by-2050-unless-urgent-action-is-taken-world-bank-report

21. For towards zero waste resources go to www.wetlandsystems.ie/wastetips.html

22. Marian Swain (2018) 'Can palm oil deforestation be stopped?' *The Breakthrough Institute*. https://thebreakthrough.org/index.php/programs/conservation-and-development/can-palm-oil-deforestation-be-stopped

23. Arthur Neslen, 1 Feb 2018, 'Nestlé under fire for marketing claims on baby milk formulas'. *The Guardian*. www.theguardian.com/business/2018/feb/01/nestle-under-fire-for-marketing-claims-on-baby-milk-formulas

24. BDS (site visited Jan 2019) 'What is BDS?' *BDS*. http://bdsmovement.net/what-is-bds

25. Nathalie Remy, Eveline Speelman and Steven Swartz, Oct 2016, 'Style that's sustainable: A new fast-fashion formula'. *McKinsey & Company*. www.mckinsey.com/business-functions/sustainability/our-insights/style-thats-sustainable-a-new-fast-fashion-formula

26. Elgin D (1998) *Voluntary Simplicity – Towards a way of life that is outwardly simple, inwardly rich*. Harper Paperbacks. NY, USA.

27. Mike Berners-Lee and Duncan Clark, 23 Sept 2010, 'What's the carbon footprint of… a new car?' *The Guardian*. www.theguardian.com/environment/green-living-blog/2010/sep/23/carbon-footprint-new-car

28. Marc Gunther, 20 Dec 2012, 'Amazon's no show on sustainability'. *The Guardian*. www.theguardian.com/sustainable-business/amazon

29. Christina Finn, 9 Jan 2018 'China took 95% of Ireland's plastic waste – but now it's changed its mind and we're in trouble'. *Thejournal.ie*. www.thejournal.ie/ireland-plastic-waste-3786393-Jan2018

30. CIEL, May 2019, 'Plastic and Climate – the hidden cost of a plastic planet'. *Centre for International Environmental Law*. www.ciel.org/plasticandclimate

31. Sourcewatch, undated – visited on 17 May 2019, 'Fracking and climate change'. *Center for Media and Democracy*. www.sourcewatch.org/index.php/Fracking_and_climate_change

32. www.communitywoodrecycling.org.uk/stores

33. www.breastcanceruk.org.uk/reduce-your-risk/back-to-school/healthier_lunches
34. Project Drawdown (2014-2019) 'Food – Reduced Food Waste'. *Drawdown*. www.drawdown.org/solutions/food/reduced-food-waste
35. Tree, I. (2018) *Wilding*. Pan McMillan, London.
36. Friends of the Earth (2006) Briefing – Genetically modified animal feed. Friends of the Earth, London.
37. Harriet Grant, 15 Dec 2018 'Legal plastic content in animal feed could harm human health, experts warn. *The Guardian*. www.theguardian.com/environment/2018/dec/15/legal-plastic-content-in-animal-feed-could-harm-human-health-experts-warn
38. Natalie Butler, 5 March 2018, 'The truth about aspartame side effects. *Healthline*. www.healthline.com/health/aspartame-side-effects
39. Rebecca O'Connell, 3 Feb 2015. 'Iceland's Last McDonald's Burger Finally Removed from Museum.' *Mental Floss*. http://mentalfloss.com/article/61439/icelands-last-mcdonalds-burger-finally-moved-national-museum
40. Maia Appleby, 27 Dec 2018, 'What are the benefits of eating multiple colored fruits and vegetables'. *Healthy Eating / SF Gate*. https://healthyeating.sfgate.com/benefits-eating-multiple-colored-fruits-vegetables-4676.html
41. Sharon Palmer, 31 March 2014, 'Detecting adulteration in olive oil'. *Food Quality and Safety*. www.foodqualityandsafety.com/article/detecting-adulteration-in-olive-oil
42. Joseph Mercola, undated, 'Fats: Why healthy dietary fat is crucial'. *Mercola*. www.mercola.com/nutritionplan/beginner_fats.htm
43. Planck N. (2016) *Real Food: what to eat and why*. Bloomsbury, NY, USA.
44. Joseph Mercola, 30 December 2013, 'The health benefits of consuming organ meats'. *Mercola*. https://articles.mercola.com/sites/articles/archive/2013/12/30/eating-organ-meats.aspx
45. Hobday, R. (1999) *The Healing Sun – sunlight and health in the 21st century*. Findhorn Press, Scotland.
46. Kat Eschner, 24 Feb 2017, 'You can still buy pig-hair toothbrushes'. *Smithsonian*. www.smithsonianmag.com/smart-news/you-can-still-buy-pig-hair-toothbrushes-180962216
47. Jessica Aldred, 19 April 2016, 'Microplastics: which beauty brands are safe to use?' *The Guardian*. www.theguardian.com/environment/2016/apr/19/microplastics-which-beauty-brands-are-safe-to-use
48. Nagel, R. (2011) *Cure Tooth Decay Naturally*. Golden Child Publishing. Ca, USA.
49. Natalie Fee, 19 Dec 2017, 'Why plastic pollution is personal'. *TEDxBristol*. www.youtube.com/watch?v=zJiQt9ASSNg
50. Martin Armstrong, 5 Oct 2018, 'The U.S. leads the world in toilet paper consumption' *Statista*. www.statista.com/chart/15676/cmo-toilet-paper-consumption
51. Susmita Baral, 23 May 2017, 'Bidets can save 15 million trees annually – so why aren't we using them?' *Life Hacker*. https://lifehacker.com/bidets-can-save-15-million-trees-annually-so-why-arent-1795431390
52. 'DIY Beauty and Natural Cleaning'. *Going Zero Waste*. www.goingzerowaste.com/beauty
53. Hawken P. (2017) *Drawdown: the most comprehensive plan ever to reverse global warming*. Penguin Books, London, UK.

54. Deborah Drew and Genevieve Yehounme, 5 July 2017, 'The apparel industry's environmental impact in 6 graphics'. *World Resources Institute*. www.wri.org/blog/2017/07/apparel-industrys-environmental-impact-6-graphics

55. Leon Kaye, 23 June 2011, 'Textile recycling innovation challenges clothing industry'. *The Guardian*. www.theguardian.com/sustainable-business/textile-recycling-challenges-industry

56. Rebecca Smithers, 11 July 2012, 'Unused clothing worth £30bn, report finds. *The Guardian*. www.theguardian.com/environment/2012/jul/11/unused-clothing-wardrobe

57. Chevalier G., Sinatra S.T., Oschman J.L., Sokal K. and Sokal P. (2012) 'Earthing: Health Implications of Reconnecting the Human Body to the Earth's Surface Electrons'. *J Environ Public Health*. 2012; 2012: 291541.

58. Alden Wicker, 27 Aug 2018, 'Acetate Sunglasses: Eco-friendly or just greenwashing?'. *EcoCult*. https://ecocult.com/acetate-sunglasses-eco-friendly-greenwashing-sustainable

59. Lucy Siegle, 2 May 2010, 'Can I wear glasses or contacts and still be green?' *The Guardian*. https://www.theguardian.com/environment/2010/may/02/lucy-siegle-wear-glasses-contacts-3d

60. Brofman M. (2004) *Improve Your Vision*. Findhorn Press, UK.

61. Corrinne Burns, 8 Aug 2014 'What happens to the excreted drugs you flush down the toilet?' *The Guardian*. www.theguardian.com/science/blog/2014/aug/08/drugs-toilet-pharmaceutical-pollution

62. Brian Clowes, undated, 'What are the environmental impacts of hormonal birth control'. *Human Life International*. www.hli.org/resources/what-are-the-environmental-impacts-of-hormonal-birth-control/#_edn9

63. Robin McKie, 2 June 2012, '£30bn bill to purify water system after toxic impact of contraceptive pill'. *The Guardian*. www.theguardian.com/environment/2012/jun /02/water-system-toxic- contraceptive-pill

64. Fiona Macdonald, 20 April 2017, 'Study finds the birth control pill has a pretty terrible impact on women's wellbeing'. *Science Alert*. www.sciencealert.com/major-study-finds-the-pill-has-a-pretty-crap-impact-on-women-s-wellbeing

65. Liz Lang (2019) '11 un-sexy chemicals lurking in condoms'. *Paleohacks*. https:// blog.paleohacks.com/chemicals-condoms

66. John Harris, 17 July 2018, 'Our phones and gadgets are now endangering the planet'. *The Guardian*. www.theguardian.com/commentisfree/2018/jul/17/internet-climate-carbon-footprint-data-centres

67. Project Drawdown (2014-2019) 'Solutions – summary of solutions by overall rank'. *Drawdown*. www.drawdown.org/solutions-summary-by-rank

68. Joseph Mercola, 23 Oct 2016, 'How LED lighting may compromise your health'. *Mercola*. https://articles.mercola.com/sites/articles/archive/2016/10/23/near-infra-red-led-lighting.aspx

69. Ben Coxworth, 14 Feb 2011, 'LED bulbs not as eco-friendly as some might think'. *New Atlas*. https://newatlas.com/led-bulbs- found-to-contain-toxic-metals/17876

70. Harvard Health Letter (updated Aug 2018) 'Blue light has a dark side'. *Harvard Health Publishing*. www.health.harvard.edu/staying-healthy/blue-light-has-a-dark-side

71. Arthur Nelson, 23 Aug 2018, 'Europe to ban halogen lightbulbs'. *The Guardian*. www.theguardian.com/environment/2018/aug/23/europe-to-ban-halogen-lightbulbs

72. Barbara Kyle, 16 Jan 2013, 'New research shows CFLs and LED light bulbs have higher toxicity and resource depletion than incandescent bulbs'. *Electronics TakeBack Coalition*. www.electronicstakeback. com/2013/01/16/new-research-shows-cfls-and-led-lightbulbs-have-higher-toxicity-and-resource-depletion -than-incandescent

73. Rachel Z. Arndt, 18 Nov 2014, 'The world's longest-burning light bulb has shone for 110 years'. *Popular Mechanics*. www.popularmechanics. com/technology/a13220/the-worlds-longest-burning-light-bulb-has-shone-for-110-years-17441176

74. Adam Hadhazy, 12 June 2016, 'Here's the truth about the 'planned obsolescence' of tech'. *BBC*. www. bbc.com/future/story/20160612-heres-the-truth-about-the-planned-obsolescence-of-tech

75. Hall K. and Nicholls R. (2008) *The Green Building Bible*. Llandysul, Green Buildling Press, Carmarthen-shire, UK.

76. Whitefield P. (2016) *The Earth Care Manual: a permaculture handbook for Britain and other temperate climates*. Permanent Publications, Hampshire, UK.

77. Seymour J. (1997) *The Complete Book of Self Sufficiency*. Dorling Kindersley, London, UK.

78. USEPA, undated, site visited 16/5/19, 'Understanding Global Warming Potentials'. *United States Environmental Protection Agency*. www.epa.gov/ghgemissions/under-standing-global-warming-potentials

79. Seth Wynes and Kimberley A Nicholas, 12 July 2017, 'The climate mitigation gap: education and government recommendations miss the most effective individual actions', *Environ-mental Research Letters*. https://

iopscience.iop.org/article /10.1088/1748-9326/aa7541

80. Project Drawdown (2014-2019) 'Solutions – summary of solutions by overall rank'. *Drawdown*. www.draw down.org/solutions-summary-by-rank

81. Tobias Long, 13 April 2017, 'The controversial third ethic of perma-culture'. *Permaculture Research Institute*. https://permaculturenews. org/2017/04/13/controver-sial-third-ethic-permaculture

82. Hans Rosling, June 2010, 'Global population growth, box by box'. *TED@Cannes*. www.ted.com/talks/ hans_rosling_on_global_popula-tion_growth

83. Stephany Molenko Baughman, 16 Jan 2018, 'Aluminum foil, how toxic is it really?' *Medium Corporation*. https:// medium.com/fryegg/aluminum-foil-how -toxic-is-it-really-2eae93cd cf97

84. Ken Robinson, Feb 2006, 'Do schools kill creativity?' *TED2006*. www.ted.com/talks/ken_robinson_ says_schools_kill_creativity

85. Saimi Jeong, 11 Nov 2018, 'Pushing up trees: is natural burial the answer to crowded cemeteries?' *The Guardian*. www.theguardian. com/lifeandstyle/2018/nov/11/ pushing -up-trees-is-natural-burial-the -answer-to-crowded-cemeteries

86. Rity Prasad, 30 Jan 2019, 'How do you compost a human body – and why would you?' *BBC News*. www. bbc.com/news/world-us-canada -47031816

87. Jonathan Fincher, 3 May 2013, 'L'Uritonnoir puts festival-goers' urine to good use'. *New Atlas*. https:// newatlas.com/luritonnoir-urine-into -fertilizer/27370

88. Top 5 Reviews, updated 25 Feb 2019, 'The 5 best female urinals [ranked]'. *Top5Reviewed.com*. www. top5reviewed.com/female-urinals

89. Anna Piccoli, visited May 2019, 'This is lady p – the female urinal created by Marian Loth'. *Mediamatic.net*. www.mediamatic.net/en/page/234343/this-is-lady-p

90. Whitefield P. (2000), *Permaculture in a Nutshell*, (2002) *How to Make a Forest Garden*. Permanent Publications, Hampshire, UK.

91. Directory of tool libraries: http://localtools.org/find

92. John Abraham, 7 Aug 2017, 'Fossil fuel subsidies are a staggering $5 tn per year'. *The Guardian*. www.theguardian.com/environment/climate-consensus-97-per-cent/2017/aug/07/fossil-fuel-subsidies-are-a-staggering-5-tn-per-year

93. Adam Conner-Simons, 4 Jan 2017, 'How ride-sharing can improve traffic, save money, and help the environment.' *MIT News*. https://news.mit.edu/2016/how-ride-sharing-can-improve-traffic-save-money-and-help-environment-0104

94. Air BnB, 2019, 'New Study Reveals A Greener Way to Travel: Airbnb Community Shows Environmental Benefits of Home Sharing'. *Air BnB*. www.airbnb.ie/press/news/new-study-reveals-a-greener-way-to-travel-airbnb-community-shows-environmental-benefits-of-home-sharing

95. Elizabeth Jardim, 26 Feb 2017, 'From smart to senseless; The global impact of ten years of smartphones'. *Greenpeace*. www.greenpeace.org/usa/research/from-smart-to-senseless-the-global-impact-of-ten-years-of-smartphones

96. Don Comis, 5 Feb 2003, 'Glomalin: the Real Soil Builder'. *USDA*. www.ars.usda.gov/news-events/news/research-news/2003/glomalin-the-real-soil-builder

97. Shemina Davis, 2 Dec 2015, 'Soil loss: an unfolding global disaster'. *University of Sheffield*. www.sheffield.ac.uk/news/nr/soil-loss-climate-change-food-security-sheffield-university-1.530115

98. Toensmeier E. (2019) *The Carbon Farming Solution*. Chelsea Green Publishing, VT USA.

99. Appelhof M. (1982) *Worms Eat My Garbage*, Flower Press, Mich., USA.

100. www.ipcc.ie/advice/composting-diy/composting-using-a-wormery

101. Worm Composting – 4 easy steps to successful worm composting. *Christchurch City Council*. www.ccc.govt.nz/assets/Documents/Environment/Sustainability/CompostWormFarm.pdf

102. DAF – Queensland Government, undated site visited 23 May 2019, 'Worm farming with piggery solid waste'. *Queensland Government – Department of Agriculture and Fisheries*. www.daf.qld.gov.au/business-priorities/agriculture/animals/pigs/managing-environmental-impacts/worm-farming

103. Álvarez-Solís J, JA Mendoza-Núñez, NS León-Martínez, J Castellanos-Albores, FA Guitérrez-Miceli (2016) 'Effect of bokashi and vermicompost leachate on yield and quality of pepper (*Capsicum annuum*) and onion (*Allium cepa*) under monoculture and intercropping cultures'. *Cien. Inv. Agr.* 43(2):243-252.

104. Alys Fowler, 22 Nov 2014, 'Alys Fowler: Hot compost bin'. *The Guardian*. www.theguardian.com/lifeandstyle/2014/nov/22/alys-fowler-hotbin

105. William Grimes, 13 Aug 1992, 'Seeking the truth in refuse', *New York Times*. www.nytimes.com/1992/08/13/nyregion/seeking-the-truth-in-refuse.html

106. Matthew Cimitile, 6 Feb 2009, 'Is coal ash in soil a good idea?' *Scientific American*. www.scientificamerican.com/article/coal-ash-in-soil

107. Stacy Sackett, 14 March 2016, 'Is newspaper safe for your garden?' *Permaculture News.* https://perma culturenews.org/2016/03/14/is-newspaper-safe-for-your-garden

108. 1001Pallets, undated – visited 17 May 2019, 'How to tell if a wood pallet is safe for reuse?' *1001Pallets.* www.1001pallets.com/pallet-safety

109. Jenkins J. (2005) *Humanure Handbook – a guide to composting human manure.* Jenkins Publishing, Pennsylvania, USA.

110. ORAU, last updated 20 Jan 2009, 'Glossy Paper'. *Oak Ridge Associated Universities.* www.orau.org/PTP/collection/consumer%20products/magazines.htm

111. Adagunodo T.A., George A.I., Ojoawo I.A., Ojesanmi K. and Ravisankar R. 'Radioactivity and radiological hazards from a kaolin mining field in Ifonyintedo, Nigeria'. *MethodsX* 5 (2018) 362- 374.

112. Udeajah R.A. (2013) 'Ecological Impact of Paper Production: A Case for the Abolition of Print Media'. *Academic Journal of Interdisciplinary Studies*, Vol 2, No.13 pp139-148.

113. EPN Staff, 24 Sept 2014, 'The environmental impacts of using paper'. *Environmental Professionals Network.* http://environmentalprofessionals network.com/the-environmental-impacts-of-using-paper

114. Hann S., Sherrington C., Jamieson O., Hickman M., Kershaw P., Bapasola A. and Cole G. (2018) *Investigating options for reducing releases in the aquatic environment of microplastics emitted by (but not intentionally added in) products – Final report.* Eunomia, Bristol, UK.

115. https://en.wikipedia.org/wiki/List_of_refrigerants

116. Oil Change International, undated – visited 17 May 2019, 'Fossil fuel subsidies overview'. *Oil Change International.* http://priceofoil.org/fossil-fuel-subsidies

117. Johnson M., Harfoot M., Muser C., Wiley T., Pollitt H., Cheqpreecha U. and Tarafdar J. (2008) *A Study in Personal Carbon Allocation: Cap and Share.* Comhar – Sustainable Development Council, Dublin, Ireland.

118. Ian Traynor, 8 May 2014, '30,000 lobbyists and counting: is Brussels under corporate sway?' *The Guardian.* www.theguardian.com/world/2014/may/08/lobbyists-european-parliament-brussels-corporate

119. Morgan J. and Mitchell P. (2015) *Employment and the circular economy – Job creation in a more resource efficient Britain.* WRAP, Oxon and Green Alliance, London, UK.

120. Jules Peck, 29 April 2013, 'The future of business: what are the alternatives to capitalism?' *The Guardian.* www.theguardian.com/sustainable-business/future-business-alternatives-capitalism

121. Klein N. (2014) *This Changes Everything: Capitalism vs. the climate.* Simon and Schuster, NY, USA.

122. American Petroleum Institute, undated, 'Who owns the oil companies', *American Petroleum Institute.* https://whoownsbigoil.com/#/?section=whoowns-the-oil-companies-2

123. ISA, undated, 'Guide to World Social Indices'. *International Sociological Association.* www.isa-sociology.org/en/opportunities/world-social-indices

124. 'Positive Money – How banks create money'. *Positive Money Europe.* https://positivemoney.org/how-money-%20works/how-banks-%20create-money

125. Benjamin Franta, 19 Sept 2018, 'Shell and Exxon's secret 1980s climate change warnings.' *The Guardian.* www.theguardian.com/environment/climate-consensus-

97-per-cent/2018/sep/19/shell-and-exxons-secret-1980s-climate-change-warnings

126. Bradford Plumer, 22 May 2006, 'The Origins of Anti-Litter Campaigns.' *Mother Jones*. www.motherjones.com/politics/2006/05/origins-anti-litter-campaigns

127. Suzi Gage, 21 Oct 2013, 'BMJ to ban research funded by the tobacco industry'. *The Guardian*. www.theguardian.com/science/sifting-the-evidence/2013/oct/21/medical-research-health

128. Union of Concerned Scientists (2012), *Heads they win, tails we lose – how corporations corrupt science at the public's expense*. USC, Cambridge, MA, USA.

129. Common Cause Foundation (2016) *Perceptions Matter: The Common Cause UK Values Survey*. Common Cause Foundation, London, UK.

130. Mollison B. (1988) *Permaculture: A Designers' Manual*. Tagari Publications, Tyalgum, Australia.

131. Mollison B. and Slay R.M. (1991) *Introduction to Permaculture*. Tagari Publications, Tyalgum, Australia.

132. www.permaculture.org.uk/knowledge-base/principles

133. Holmgren D. (2011) *Permaculture – Principles and Pathways Beyond Sustainability*. Permanent Publications, Hampshire, UK.

134. www.permaculture.co.uk/articles/what-permaculture-part-1-ethics

135. 'Women get the vote'. *Parliament. uk* www.parliament.uk/about/living-heritage/transformingsociety/elections voting/womenvote/overview/thevote

136. Avinash Bhunjun, 6 Feb 2018 'When did all men get the vote and when did the voting age change to 18?' *Metro*. https://metro.co.uk/2018/02/06/when-did-all-men-get-the-vote-and-when-did-the-voting-age-change-to-18-7290759

137. Undated, site visited 21 May 2019 'Who got the right to vote when? A history of voting rights in America' *Aljazeera*. https://interactive.aljazeera.com/aje/2016/us-elections-2016-who-can-vote/index.html

138. Undated, site visited 21 May 2019 'Fast overview on segregation'. *Laws*. https://civil.laws.com/segregation

139. First Peoples Worldwide, 2 Sept 2013, '3 horrendous anit-indigenous laws'. *Cultural Survival*. www.culturalsurvival.org/news/3-horrendous-anti-indigenous-laws

140. Marie O'Halloran, 28 June 2017, 'Ireland joins France, Germany and Bulgaria in banning fracking'. *Irish Times*. www.irishtimes.com/news/politics/oireachtas/ireland-joins-france-germany-and-bulgaria-in-banning-fracking-1.3137095

141. Damian Carrington, 12 July 2018, 'Ireland becomes world's first country to divest from fossil fuels.' *The Guardian*. www.theguardian.com/environment/ 2018/jul/12/ireland-becomes-worlds-first-country-to-divest-from-fossil-fuels

142. City to Sea, 7 Dec 2016, 'Retailers pledge to 'switch the stick' to stop source of plastic pollution'. *City to Sea*. www.citytosea.org.uk/seven-major-retailers-pledge-to-switch-the-stick-to-stop-source-of-plastic-pollution

143. Damian Carrington, 5 July 2017, 'G20 public finance for fossil fuels 'is four times more than renewables', *The Guardian*. www.theguardian.com/world/2017/jul/05/g20-public-finance-for-fossil-fuels-is-four-times-more-than-renewables

144. Common Cause Foundation (2016) *Perceptions Matter: The Common Cause UK Values Survey*. Common Cause Foundation, London, UK.

145. Personal communication

146. Sam Sholli, 11 Dec 2018 'Old tyres used for road construction'. *New Civil Engineer*. www.newcivilengineer.com/latest/old-tyres-used-for-road-construction/10038087.article

147. 'Waste – Construction and Demolition Waste (CDW)'. *European Commission*. http://ec.europa.eu/environment/waste/construction_demolition.htm

148. Nav.sparx, undated, visited 21 May 2019, 'How to construct a house with plastic bottles!!' *Instructables*. www.instructables.com/id/New-Innovation-in-Construction-using-Waste-Plastic

149. David Matthews, 7 May 2010, 'Earthship: Sustainable building with 900 spare tyres'. *Building*. www.building.co.uk/focus/earthship-sustainable-building-with-900-spare-tyres/3163060.article

150. Duuvy Jester 2 May 2014, The Multi Bioregion Earthship. Permaculture News. https://permaculturenews.org/2014/ 05/02/multi-bioregion-earthship

151. Angelique Chrisafis, 21 May 2015, 'France to force big supermarkets to give unsold food to charities', *The Guardian*. www.theguardian.com/world/2015/may/22/france-to-force-big-supermarkets-to-give-away-unsold-food-to-charity

152. Gleaner, 5 Feb 2018, 'CAPRI – Reducing 'scandal' bag use in Jamaica: ban or fee? – Denmark first country to tax plastic bags'. *Gleaner*. http://jamaica-gleaner.com/article/news/20180205/capri-reducing-scandal-bag-use-jamaica-ban-or-fee-denmark-first-country-tax

153. Jane Onyanga-Omera, 14 Sept 2013, 'Plastic bag backlash gains momentum'. *BBC News*. www.bbc.com/news/uk-24090603

154. BigFatBags, undated – visited 21 May 2019. 'List by country; bag charges, taxes and bans'. *BigFatBags*. www.bigfatbags.co.uk/bans-taxes-charges-plastic-bags

155. Lorraine Chow, 15 Feb 2018, 'China reassigns 60,000 soldiers to plant trees'. *EcoWatch*. www.ecowatch.com/china-trees-soldiers-2534965590.html

156. Brian Clark Howard, 18 July 2016, 'India Plants 50 Million Trees in One Day, Smashing World Record'. *National Geographic*. https://news.nationalgeographic.com/2016/07/india-plants-50-million-trees-uttar-pradesh-reforestation

157. Soo Youn, 17 Oct 2017, 'Visit the world's only carbon-negative country'. *National Geographic*. www.nationalgeographic.com/travel/destinations/asia/bhutan/carbon-negative-country-sustainability

158. Bryant Rousseau, 13 July 2016, 'In New Zealand, lands and rivers can be people (legally speaking)'. *New York Times*. www.nytimes.com/2016/07/14/world/what-in-the-world/in-new-zealand-lands-and-rivers-can-be-people-legally-speaking.html

159. John Vidal, 10 Apr 2011, 'Bolivia enshrines natural world's rights with equal status for Mother Earth'. *The Guardian*. www.theguardian.com/environment/2011/apr/10/bolivia-enshrines-natural-worlds-rights

160. http://theoneplanetlife.com/what-is-one-planet-development-in-wales

Index

washing
  machines 69, 79, 80,
  124-5
  powder 69
  windows 65
washing-up liquid 33, 64
waste
  food 56-7
  electronic equipment 12,
  18-9, 46, 73
  embodied 20, 22-3, 30,
  66, 71, 125, 128
  generation 32, 43, 54,
  126, 139
  hazardous 20, 25-6,
  94-5, 128-9
  management 15, 126
  minimisation 3, 12, 25,
  34-5, 39
  stream 17-9
  to energy 10, 16-7, 43, 50
  wood 45, 92-3
water 8-9
  drinking 8, 26, 66, 72
  footprint 13, 20, 24, 68,
  70-1
  pollution 26, 60, 63, 83,
  85
waterproof
  clothing 17
  covers 81
wealth 1, 7, 13, 24, 133,
  139, 142
weather 11, 31
weddings 85
weeds 104-5, 116-7
  killers 26
  suppressor 79, 117
  tea liquid fertiliser 117
wellbeing 12, 138
West Cork 89, 93
white goods 19, 80, 129
wholefood diet 60
wholesale 21, 57, 89
wild 7, 11, 90
wilding (see rewild)
wildlife
  impacts 9, 13, 22-5, 65,
  69, 83, 96
  supporting 90, 92, 114,
  121
willow 17, 85, 89, 121
windows
  PVC 44
  washing 65, 70
wipes 67
wood (see timber)
  ash 77-8, 91, 117
  clean 29, 45, 50, 118
  composites 29, 45,
  92-3, 117-8
  harvest 77, 123
  preservatives 39, 90, 94,
  117

treated 29, 45, 90, 92,
  107, 117-8
  untreated 79, 92, 107
woodchips 78, 115, 117
woodland habitats 68
woodlands 11, 92
wool 71, 89, 106, 124, 127
work as rent 97
working remotely 100
workshops 14, 145
workspaces 100
worms 88, 106-10, 112-4,
  139
  brandling 108
wrappers 7, 34, 67, 87, 123
WWOOFing 97

yoghurt pots 36, 38, 54, 58,
  111
young people 74, 142

zero impact 50
zero waste 3-6, 28, 47-8,
  53, 112
  shops 14, 130
  society 14, 16, 47,
  126-7
zooplankton 69

Enjoyed this book? You may also like these from Permanent Publications

Edible Paradise	**Vital Skincare**	**The Creative Kitchen**
Vera Greutink	Laura Pardoe	Stephanie Hafferty
£16.00	£19.95	£19.95

Create your own no dig, organic garden with permaculture design and techniques, including herbs, vegetables and flowers, for a beautiful and productive garden.

Over 100 Blend-It-Yourself skincare recipes using hedgerow and garden herbs to make your own face creams, moisturisers, shampoos, toners, cleansers and more.

How to make seasonal, plant-based meals, drinks, soaps, balms, and store cupboard ingredients like vinegars and essences from local, homegrown produce.

Our titles cover: permaculture, home & garden, green building, food & drink, sustainable technology, woodlands, community, wellbeing and so much more

Available from all good bookshops and online retailers, including the publisher's online shop:
https://shop.permaculture.co.uk
with 10% off the RRP on all books

Our books are also available via our American distributor, Chelsea Green:
www.chelseagreen.com/publisher/permanent-publications

Permanent Publications also publishes *Permaculture Magazine*